One-Liners on Soil and Water Management

NIPA GENX ELECTRONIC RESOURCES & SOLUTIONS P. LTD.
New Delhi-110 034

About the Authors

Er. Abhishek Patel is a scientist under Land and Water Management Engineering section of ICAR- Central Arid Zone Research Institute, RRS-Bhuj, Gujarat. As per his educational qualifications, he had completed B.Tech (Agricultural Engineering) from Dr. Panjabrao Deshmukh Krishi Vidyapeeth, Akola (Maharashtra), M. Tech in Land and Water Resource Engineering from Department of Agricultural & Food Engineering, Indian Institute of Technology-Kharagpur. His expertise includes remote sensing and GIS, hydrological modelling etc. He is currently dealing with grassland development and water management in arid ecosystem.

Dr. Anandkumar Naorem is a soil scientist and currently working as Head (I/C), ICAR-Central Arid Zone Research Institute, RRS-Bhuj, Gujarat. He completed his doctoral research from Bidhan Chandra Krishi Viswavidyalaya, West Bengal, master's research from College of Post-Graduate Studies, Central Agricultural University, Meghalaya and B.Sc. (Agriculture) from Assam Agricultural University, Jorhat. His expertise and specialization include soil carbon sequestration, heavy metal remediation, bioformulation, fodder cactus and land-use change study. He is currently dealing with both international and institute projects related to the dryland ecosystem and its associated soil qualities. The scientific literature repository of Dr. A.K. Naorem includes soil related changes under conservation agriculture, introduction of spineless cactus in arid conditions, immobilizing agents in heavy metal toxicity in soils, concerns of dryland grasslands etc.

One-Liners on Soil and Water Management

Abhishek Patel
Scientist (Land & Water Management Engineering)
ICAR - Central Arid Zone Research Institute
Regional Research Station Bhuj, Gujarat - 370 105

Anandkumar Naorem
Scientist (Soil Science)
ICAR - Central Arid Zone Research Institute
Jodhpur, Rajasthan - 342 003

NIPA GENX ELECTRONIC RESOURCES & SOLUTIONS P. LTD.
New Delhi-110 034

NIPA GENX ELECTRONIC RESOURCES & SOLUTIONS P. LTD.

101,103, Vikas Surya Plaza, CU Block
L.S.C.Market, Pitam Pura, New Delhi-110 034
Ph : +91 11 27341616, 27341717, 27341718
E-mail:newindiapublishingagency@gmail.com
www: www.nipabooks.com

For customer assistance, please contact
Phone: + 91-11-27 34 17 17 Fax: + 91-11- 27 34 16 16
E-Mail: feedbacks@nipabooks.com

ISBN: 978-93-94490-55-0

Composed and Designed by NIPA.

Dedicated to

Our Parents

(***Vishnu Prasad Patel, Rajkumari Patel***)

(***Naorem Jyotinkumar, Asem Sobita Devi***)

Preface

The desire for publishing this handy book has originally come from our long-time appetite for writing a non-voluminous book on soil and water management that can be used by students or any readers who need to have a quick glance on the scientific jargons, that are mostly used in soil and water management studies. Both the authors prepared intensely to crack ARS examination in soil sciences and land and water management engineering and are currently constituting the scientific team of ICAR-Central Arid Zone Research Institute, RRS-Bhuj. During the course of their journey, one of the most crucial yet a time-consuming effort was to make a note on the definitions and simplified explanations of the scientific terms in soil and water management. It's never too late to address the issues we all were a part of. The book aims to constructively write a handy compilation of all these scientific terms along with their simplified definitions so that the readers are offered with a tailored note of one-liners from each respective sub-topics.

This book wouldn't have been possible without the support and technical help of our official team of ICAR-Central Arid Zone Research Institute, RRS-Bhuj. We owe an enormous debt of gratitude to Sh. Mahaveer Singh Rathore, T-I, ICAR-CAZRI, RRS-Bhuj for converting the long hand-written notes to typed forms. Thanks to our parents for never losing hope in us and ultimately all our friends who always encouraged us.

Authors

Contents

Syllabus

Land & Water Management Engineering

ARS Subject Code: 53

Unit 1: Groundwater Development, Wells and Pumps

Water resources of India. Irrigation potential and contribution of groundwater, utilizable groundwater resources and level of groundwater development in the country, scope of groundwater development. Aquifer types and parameters. Principles of groundwater flow, interaction between surface and groundwater, natural and artificial groundwater recharge. Hydraulics of fully and partially penetrating wells. Design, construction and development of irrigation wells. Water lifts, pumps and prime movers, well and pumps characteristics, performance evaluation and selection of pumps. Energy requirement in groundwater pumping. Design of centrifugal pumps. Groundwater pollution. Salt water intrusion in inland and coastal aquifers. Application of groundwater models for groundwater development and management. Conjunctive use of surface and groundwater.

Unit 2: Open Channel Hydraulics

Hydraulics of open channel flow, energy and momentum principles, specific energy, Hydraulic jump and its classification. Design of different types of irrigation channels. Irrigation water measurement: using velocity area method, water meters, weirs, notches, flumes, orifices etc. Water conveyance and control. Conveyance losses and lining of irrigation channels. Irrigation water delivery and distribution.

Unit 3: Soil, Plant, Water and Atmosphere Relationship

Soil and water as vital resources for agricultural production. Water retention by soil, soil moisture characteristics, field capacity, permanent wilting point, plant available water and extractable water. Soil irritability classifications, factors affecting profile water storage. Determination of soil water content, computation of soil water depletion, soil water potential and its components, hydraulic head. Field water budget water gains and water losses from soil, deep percolation beyond root zone, capillary rise. Evapotranspiration (ET) and irrigation requirement, critical stages of crop growth in relation to irrigation. Irrigation scheduling. Plant water relations, concept of plant water potential, significance of osmotic adjustment, leaf diffusive resistance, canopy temperature, canopy temperature depression (CTD). Water movement through soil plant atmosphere system. Uptake and transport of water by roots. Development of crop water deficit, crop adaptation to water deficit, morpho physiological effect of water deficit. Drought tolerance, mechanisms of drought tolerance. potential drought tolerance traits and their measurements. Management strategies to improve crop productivity under limited water supplies. Contingent crop plans and other strategies for aberrant weather conditions. Cropping patterns, alternate land use and crop diversification in rain fed regions.

Unit 4: Hydrology and Soil and Water Conservation

Hydrologic cycle, precipitation, infiltration and surface runoff. Measurement and analysis of hydrologic data. Application of statistics in hydrology. Probability concepts. Distributions and application. Intensity duration frequency analysis. Hortonian and saturation overland flow theories, partial source area concept of surface runoff generation. Rainfall and run off relationships, stream gauging and runoff measurement. Different methods of surface runoff estimation, hydrographs, S-hydrograph, IUH, unit hydrograph theory and its application, Flood routing methods and calculations. Soil erosion and types of erosion. Soil loss measurement and 128 estimations. Universal soil loss equation and subsequent its modifications, soil and water conservation structures and their design. Gully control structures and their design. Design and construction of farm pond and reservoir. Seepage theory. Design of earthen dams and retaining walls, stability analysis of slopes. Mathematical models and simulation of hydrologic processes. Application of GIS in soil and water conservation.

Unit 5: Watershed Management

Watershed concept, Identification and characterization of watersheds. Hydrological and geomorphological characteristics of watersheds. Land capability and irritability classification and soil maps. Principles of watershed management. Development of watershed management plans, its feasibility and economic evaluation. Land levelling and grading, machineries and equipments for land levelling.

Unit 6: Irrigation Water Management

History of irrigation in India. Management of irrigation water. Major irrigation projects in India. Crop water requirements. Soil water depletion, plant indices and climatic parameters. Crop modeling, water production function. Methods of irrigation, surface methods, overhead methods, Pressurized irrigation system such as drip and sprinkler irrigation. Merits and demerits of various methods. Hydraulics of furrow, check basin and border irrigation, Hydraulics and design of pressurized irrigation systems. Irrigation efficiency and economics of different irrigation systems. Application and distribution efficiencies. Agronomic considerations in the design and operation of irrigation projects, characteristics of irrigation and farming systems affecting irrigation management. Irrigation legislation. Irrigation strategies under different situation of water availability, optimum crop plans and cropping patterns in canal command areas. Quality of irrigation water and irrigation with poor quality water. On farm water management, socio- economic aspects of on farm water management. Scope for economizing the use of water.

Unit 7: Management of Degraded, Waterlogged and Other Problematic Soils and Water

Problem soils and their distribution in India. Water quality criteria and use of brackish waters in agriculture. Excess salt and salt tolerant crops. Hydrological imbalances and their corrective measures. Concept of critical water table depths for crop growth. Contribution of shallow water table to crop water requirements. Management strategies for flood prone areas and crop calendar for flood affected areas. Crop production and alternate use of problematic soils and fish production. Agricultural field drainage and theory of flow in saturated soil. Flow net theory and its application. Drainage investigations. Drainage characteristics of various type of soils. Water table contour maps and isobaths maps. Drainage coefficient. Design and installation of surface and subsurface drainage system. Interceptor and relief drains and their design. Drain pipe and accessories. Pumped drainage. Drainage requirements of crops. Drainage in relation to salinity and water table control. Reclamation of ravine, waterlogged, swampy areas and polders. Salt-affected soils and their reclamation. Command area development organizational structures and activities. River valley projects, interstate disputes. Water rights and legal aspects. Irrigation water user's association concept and responsibilities. Environmental considerations in land and water resources management.

Topic wise Recommended Books for ARS syllabus

Topic(s)	Book
Hydrology	
Hydrologic cycle, precipitation, infiltration and surface runoff. Measurement & analysis of hydrologic data. Application of statistics in hydrology. Probability concepts. Distributions and application. Intensity duration frequency analysis. Hortonian and saturation overland flow theories, partial source area concept of surface runoff generation. Rainfall and run off relationships, stream gauging and runoff measurement. Different methods of surface runoff estimation, hydrographs, S-hydrograph, IUH, unit hydrograph theory and its application, Flood routing methods and calculations.	1. Watershed Hydrology (Suresh R) 2. Engineering Hydrology (Subrahmanya, K) 3. Land and Water Management (Murty V.V.N. and Jha M.K.)
Soil and Water Conservation	
Soil erosion and types of erosion. Soil loss measurement and estimation. Universal soil loss equation and subsequent its modifications, soil and water conservation structures and their design. Gully control structures and their design.	1. Soil and water Conservation Engineering (Suresh R) 2. Manual of Soil Water conservation practices (Singh G., Venkataraman C., Sastry G. and Joshi B.P.) 3. Land and Water Management (Murty V.V.N. and Jha M.K.)
Design and construction of farm pond and reservoir. Seepage theory. Design of earthen dams and retaining walls, stability analysis of slopes.	1. Irrigation Engineering and Hydraulic Structures (Garg S. K) 2. Land and Water Management (Murty V.V.N. and Jha M. K.)
Mathematical models and simulation of hydrologic processes. Application of GIS in soil and water conservation.	1. Land and Water Management (Murty V.V.N. and Jha M. K.) 2. R. Suresh
Soil, Plant, Water and Atmosphere Relationship	
Soil and water as vital resources for agricultural production. Water retention by soil, soil moisture characteristics, field capacity, permanent wilting point, plant available water and extractable water. Soil irrigability classifications, factors affecting profile water storage. Determination of soil water content, computation of soil water depletion, soil water potential and its components, hydraulic head. Field water budget water gains and water losses from soil, deep percolation beyond root zone, capillary rise. Evapotranspiration (ET) and irrigation requirement, critical stages of crop growth in relation to irrigation. Irrigation scheduling. Plant water relations, concept of plant water potential	1. Irrigation Theory and Practice (Michael A. M.) 2. Land and Water Management (Murty V.V.N. and Jha M. K.)

Significance of osmotic adjustment, leaf diffusive resistance, canopy temperature, canopy temperature depression (CTD). Water movement through soil plant atmosphere system. Uptake and transport of water by roots. Development of crop water deficit, crop adaptation to water deficit, morpho physiological effect of water deficit. Drought tolerance, mechanisms of drought tolerance. potential drought tolerance traits and their measurements.	**Internet Sources**
Management strategies to improve crop productivity under limited water supplies. Contingent crop plans and other strategies for aberrant weather conditions. Cropping patterns, alternate land use and crop diversification in rain fed regions.	**Internet Sources**

Irrigation Water Management

History of irrigation in India. Management of irrigation water. Major irrigation projects in India. Crop water requirements. Soil water depletion, plant indices and climatic parameters. Crop modeling, water production function.	1. Irrigation Theory and Practice (Michael A. M.) 2. Land and Water Management (Murty V.V.N. and Jha M. K.) 3. Irrigation Engineering and Hydraulic Structures (Garg S. K.)
Methods of irrigation, surface methods, overhead methods, Pressurized irrigation system such as drip and sprinkler irrigation. Merits and demerits of various methods. Hydraulics of furrow, check basin and border irrigation, Hydraulics and design of pressurized irrigation systems. Irrigation efficiency and economics of different irrigation systems. Application and distribution efficiencies	1. Irrigation Theory and Practice (Michael A. M.) 2. Land and Water Management (Murty V.V.N. and Jha M. K.)
Agronomic considerations in the design and operation of irrigation projects, characteristics of irrigation and farming systems affecting irrigation management. Irrigation legislation. Irrigation strategies under different situation of water availability, optimum crop plans and cropping patterns in canal command areas. Quality of irrigation water and irrigation with poor quality water. On farm water management, socio- economic aspects of on farm water management. Scope for economizing the use of water.	**Internet Sources**

Management of Degraded, Waterlogged and Other Problematic Soils and Water

Problem soils and their distribution in India. Water quality criteria and use of brackish waters in agriculture. Excess salt and salt tolerant crops. Hydrological imbalances and their corrective measures. Concept of critical water table depths for crop growth. Contribution of shallow water table to crop water	1. Irrigation Theory and Practice (Michael A. M.) 2. Land and Water Management (Murty V.V.N. and Jha M. K.) 3. Drainage Principles and Applications (2 ed.) ILRI Publi.

requirements. Management strategies for flood prone areas and crop calendar for flood affected areas. Crop production and alternate use of problematic soils and fish production. Agricultural field drainage and theory of flow in saturated soil. Flow net theory and its application. Drainage investigations. Drainage characteristics of various type of soils. Water table contour maps and isobaths maps. Drainage coefficient. Design and installation of surface and subsurface drainage system. Interceptor and relief drains and their design. Drain pipe and accessories. Pumped drainage. Drainage requirements of crops. Drainage in relation to salinity and water table control. Reclamation of ravine, waterlogged, swampy areas and polders. Salt-affected soils and their reclamation.	(Ritzema H. P.) 4. Irrigation Water Management Principles and Practice (Majumdar D. K.)
Command area development organizational structures & activities. River valley projects, interstate disputes. Water rights and legal aspects.	**Internet Sources**
Irrigation water user's association concept and responsibilities. Environmental considerations in land and water resources management.	1. Irrigation Theory and Practice (Michael A. M.) 2. Internet Sources

Open Channel Hydraulics

Hydraulics of open channel flow, energy and momentum principles, specific energy, Hydraulic jump and its classification. Design of different types of irrigation channels	1. Fluid Mechanics and Hydraulic Machines (Bansal, R.K.) 2. Land and Water Management (Murty V.V.N. and Jha M. K.) 3. Irrigation Engineering and Hydraulic Structures (Garg,S. K)
Irrigation water measurement: using velocity area method, water meters, weirs, notches, flumes, orifices etc. Water conveyance and control. Conveyance losses and lining of irrigation channels.	1. Fluid Mechanics and Hydraulic Machines, Bansal, R.K 2. Irrigation Theory and Practice (Michael A. M.)
Irrigation water delivery and distribution.	1. Irrigation Theory and Practice (Michael A. M.) 2. Irrigation Engineering and Hydraulic Structures, Garg,S. K

Groundwater Development, Wells and Pumps

Water resources of India.	1. Irrigation Theory and Practice (Michael A. M.)
Irrigation potential and contribution of groundwater, utilizable groundwater resources and level of groundwater development in the country, scope of groundwater development. Aquifer types and parameters. Principles of groundwater flow, interaction between surface and groundwater, natural and artificial groundwater recharge. Hydraulics of fully and	1. Water Well and Pump Engineering, (Michael A. M.) 2. Irrigation Theory and Practice (Michael A. M.) 3. Land and Water Management (Murty V.V.N. and Jha M. K.)

partially penetrating wells. Design, construction and development of irrigation wells. Water lifts, pumps and prime movers, well and pumps characteristics, performance evaluation and selection of pumps. Energy requirement in groundwater pumping. Design of centrifugal pumps. Groundwater	
Pollution. Salt water intrusion in inland and coastal aquifers. Application of groundwater models for groundwater development and management.	**Internet Sources**
Conjunctive use of surface and groundwater.	1. Irrigation Theory and Practice (Michael A. M.) 2. Land and Water Management (Murty V.V.N. and Jha M. K.)
Watershed Management	
Watershed concept, Identification and characterization of watersheds. Hydrological and geomorphological characteristics of watersheds. Land capability and irritability classification and soil maps.	1. Watershed Planning and Management, (Singh R.V.) 2. R Suresh 3. Land and Water Management (Murty V.V.N. and Jha M. K.)
Principles of watershed management. Development of watershed management plans, its feasibility and economic evaluation.	1. R Suresh 2. Land and Water Management (Murty V.V.N. and Jha M. K.) 3. Internet Sources
Land levelling and grading, machineries and equipment for land levelling	1. Irrigation Theory and Practice (Michael A. M.) 2. Land and Water Management (Murty V.V.N. and Jha M. K.)

1

Soil Erosion

Technical Term	Definition
Soil erosion	Transportation & deposition of soil particles from one place to other after detachment from soil mass.
Detachment limited erosion	When the eroding agents have sufficient capacity to transport more quantity of detached material than the material supplied through detachment, then erosion is called detachment limited
Transport limited	When eroding agent don't have sufficient capacity to transport whole quantity of material that is supplied through detachment, then erosion is called transport limited
Soil erodibility	It is the potential susceptibility of soil to get erode.
Erosivity	Potential ability of external force (rainfall/wind) to detach soil particle from soil mass.
Geological erosion	Refers to process of simultaneous loss and formation of soil (balance of soil loss and formation). Also termed as natural erosion or normal erosion.
Accelerated soil erosion	Includes serious detachment and loss of soil by natural process and anthropogenic activities.
Water erosion	When water acts as eroding agent in process of erosion, then it's called as water erosion
Terminal velocity	Constant velocity of a body falling from certain height when the frictional resistance of sir becomes equal to the gravitational force.
Critical slope length	Slope length of field at which soil erosion begins.
Critical inclination	Inclination of field at which sediment yield begin to form (sedimentation starts)

Sheet erosion (Inter-rill erosion)	More or less uniform removal of soil in the form of a thin a layer or sheet by flowing water is called sheet erosion.
Rill erosion	Erosion due to micro channels formed due to irregularity in slope of field which can be removed by plough/implement.
Gully erosion	Advance stage of rill erosion in which size of rill is so enlarge that can't be smoothened by ordinary tillage implement.
Universal Soil Loss Equation (USLE)	R, K, L, S, C, and P (Term "Universal" indicates that this equation considers all possible factors of soil erosion and also has general application)
Rainfall erosivity (R) factor	Rainfall erosivity is the potential of the rainfall to erode the soil.
Soil erodibility factor (K)	Soil erodibility is the ability (nonresistance) of parent soil (rock) to get eroded by eroding agent.
Slope length & steepness factor (LS)	Slope length & steepness facto is the ratio of soil loss from unit area of a site to the corresponding loss from a standard experimental plot (22.1 m long plot with a 9% slope). It combines the effects of slope length and slope gradient on the soil erosion.
Crop management factor (C)	Crop management factor is the ratio of soil loss under actual conditions to that from land under the reference (clean-tilled continuous-fallow conditions) conditions
Conservation practice factor (P)	Conservation practice factor is the ratio of soil loss with contouring and/or strip cropping to that with straight row up-and-down slope farming. This reflects the impact of support practices on erosion.
Wind erosion	Process of transportation, and deposition of eroded soil particle by the action of wind.
Wind erosion types	(a) Sweeping drift, (b)Active drift
Sweeping drift	Refers to the relatively gentle wind erosion, in this only some loose soil's particle gets into suspension

Active drift	Active drift refers to more vigorous wind erosion, it involves all three particle movements (saltation, suspension and surface creep)
Critical wind velocity for wind erosion	It is the wind velocity at which soil particles begin to move on soil surface.
Coefficient of soil resistance	Coefficient of soil resistance is the ratio of kinetic energy of wind to resistance force of the soil.
Forms of wind erosion	(a) Deflation, (b) Abrasion or (Corrasion)
(a) Deflation	It refers as a process in which the loosen soil particles are carried away from one place to other by wind. (Occurs mainly on loose rocks) OR blowing away soil particles is deflation.
(b) Abrasion or (Corrasion)	It is the process in which particle removal is caused by wind-carried particles. (Occurs on hard rocks)
Avalanching	It may be defined as increase in rate of soil erosion as wind blows farther across the field
Sand dune	The transported sand particles by wind are accumulated at some point on land surface, is called sand dunes

2

Soil Erosion Conservation

Overview of Soil Erosion Conservation Measures

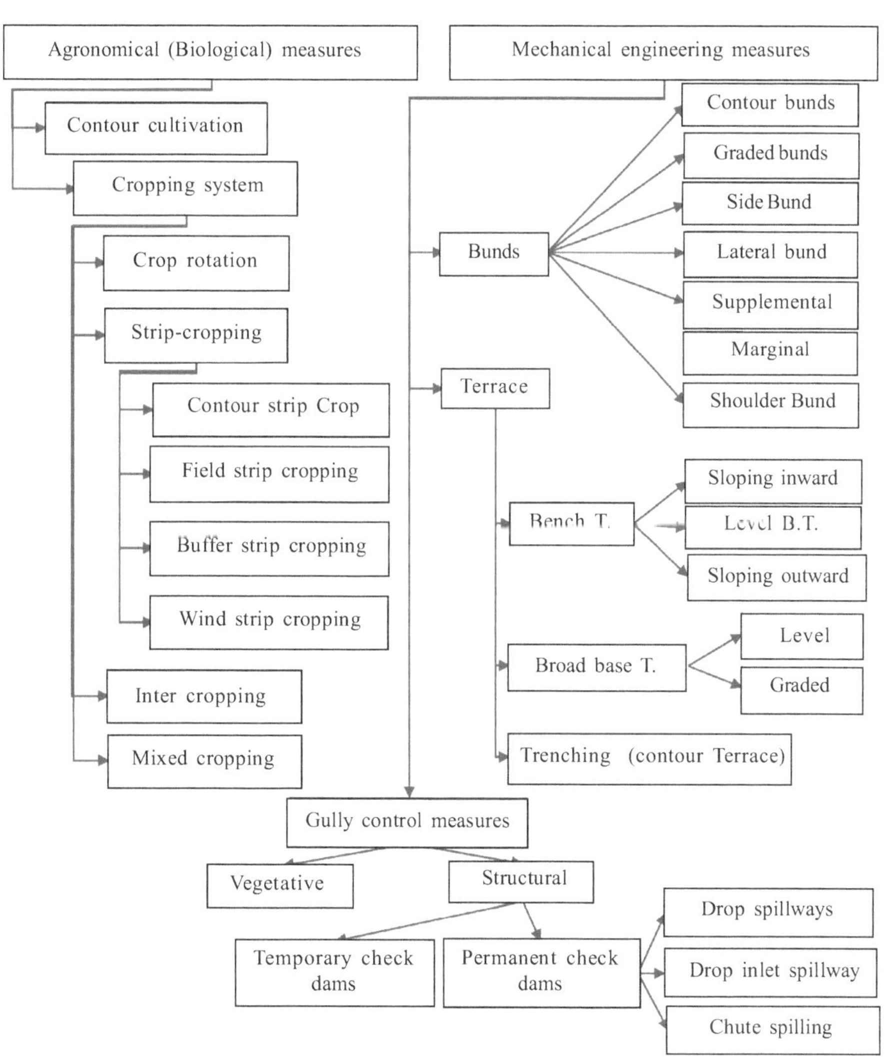

Technical Term	Definition
Conservation	Defined as protection, improvement and use of natural resources (with the condition) for maximum benefits (economics and social)
Erosion control measures	Agronomical (Biological) or mechanical (engineering) interventions intended to control soil erosion.
Agronomical (Biological) measures	Refers to agronomical practice on mild slopy lands (agricultural lands) to control soil erosion, also refers to first line of defense
1. Contour cultivation	Refers to cultivation or tillage practices performed nearly or on the contours to protect soil.
2. Cropping system	The pattern (yearly sequence & spatial arrangement) of crops growing in a given piece of land (field).
2.1 Crop rotation	Growing of selected crops of different kinds in a regular order in a given field.
2.2 Strip-cropping	It is a practice of growing of alternate strips of different types of crops in a given field.
2.2.1 Contour strip crop	Strips of crops are planted (grown) along contour across to land slope.
2.2.2 Field strip cropping	It is a modified form of contour strip cropping in which strips of crops are grown parallel to each other across land slope but not always on contour.
2.2.3 Buffer strip cropping	It is a Practice of growing strips of grass, legume or other erosion resisting vegetation between the crop's strips in field.
2.2.4 Wind strip cropping	Practice of growing strips of crop just across the prevailing wind-direction in field.
2.3 Inter cropping	Practice of growing more than one crop (base crop with associate crop) at a time in alternate rows with adequate spacing.
2.4 Mixed cropping	Practice of growing crops (two or more) at a time without any definite arrangement of row (it is in mixed pattern).
3. Conservation tillage	The tillage (mechanical manipulation of soil) practices applied to conserve the soil against erosion is called conservation tillage.

4. Mulching	Mulching is a practice of in which crop residues or other organic residues are placed on soil surface to cover it.
Double cropping	Practice of growing of two crops in a sequence each year in same field
Multiple cropping	Practice of growing of two or more crops consecutively in same field
Continuous cropping	Practice of growing of crops one after another on-same field
Conservation cropping	Method or practice of farming (growing crops) which is aimed to provide maximum protection to soil against erosion.
Relay cropping	Seedling / Planting of the succeeding crop after flowering and before harvesting of the standing crop.
Conservation Agriculture	Defined as the practice to promote infiltration of rainfall where it falls, its retention in soil, efficient use of soil moisture & nutrients for higher and more suitable production.
Mechanical (Engineering) measure	Measures for soil conservation that involve earth work and engineering construction and used for retaining water by creating obstruction, thus to control erosion.
Bund	Bunds are embankment like structure, constructed across the land slope.
Types of bunds	1. Contour bunds, 2. Graded bunds
Contour bund	Bunds constructed exactly on contour or nearly on contour are called contour bunds
Side Bund	Bunds constructed at the extreme ends of contour bund along the land slope.
Lateral bund	Bund constructed between two side bunds along the slope for preventing the concentration of water at one side. It also breaks length of slope contour bund into convenient bits
Supplemental Bund	Bund constructed between two contour bunds to limit horizontal spacing by its minimum extent.
Marginal Bund	Bund constructed at margin of the field, road, river etc. to demarcate their boundary.

Shoulder bund	Bunds constructed at outer edge of the bench terrace to held the runoff over the top off the terrace and to provide stability to the terrace.
Surplusing Arrangement in Bunds	Arrangements meant for conducting safe disposal of excess runoff from the bunded area & make bund system safe against accumulated water pressure.
Grassed Waterways	Grassed waterways are open channels protected with suitable grass used to convey water for safe dispose.
Terrace	Step like structures along with embankment or ridge, across land slope to check flow of surface water and to reduce soil loss by reducing effective length & slope of land
Bench terrace	Bench terraces are the platform like construction, which are constructed along contours of sloping land
Level bench terrace	Irrigated type bench terraces are generally constructed in irrigated conditions. (Also termed as Paddy terrace)
Bench terrace slopping outward	Bench terraces with slope outward are constructed in form of narrow strips, widely used for orchard plantation.
Bench terrace slopping inward	Bench terraces with slope inward are constructed in those hilly areas which have reverse land slope toward the hill. (Hill type bench terraces)
Broad base terrace	Broad base terrace is defined as surface channel or embankment type constructions formed across land slope. (a) Graded Terrace-Channel type terrace, constructed to drain excess water (b) Level Terrace-Constructed for conveying moisture and to control soil erosion
Trenching (contour Terrace)	Trenching is a conservation (soil & water) practice generally adopted for land capability class V-VIII due to serious limitations, not suitable for cultivation as such.
Gradonies	Gradonies are steeply inward-sloping narrow bench terrace constructed on contours

Diversion	Diversions are defined as designed channels constructed across the slope for the purpose of intercepting & diverting surface flow to a safe point.
Gully control measures	The control / conservative measures meant to control gully erosion by stabilizing the condition.
Vegetative gully control measures	It involves the growing of vegetation to provide a soil cover and protection to gully against scouring.
Gully control by structures (Check dams)	Structures are constructed and used to control the flow velocity and thereby the gully erosion called (These structures are known as check dams)
Check dams	The gully control structures constructed to check the, flowing water in gully are known as check dams
Types of check dams	1. Temporary Check dams, 2. Permanent Check dams
Temporary check dam	Temporary check dam is the structure constructed to collect soil/sediment upstream of structure to stabilize the gully via vegetation growth.
Drop spillway	Drop spillways is a permanent gully control structure used at gully bed to create a control point.
Drop inlet spillway	Drop inlet spillway is a permanent gully control structure consists of a pipe (conduit) passing through an earth fill dam with suitable inlet & outlet component.
Chute spilling	Chute spilling is an open channel like structure constructed on steep slope of gully face or head with a suitable inlet & outlet.
Spillways	Spillway is a flow path constructed either in a dam section or in a drop structure for effective and safe disposal of water from upstream to downstream end.
Emergency spillways (Auxiliary spillway)	Spillways provided to protect the embankment from overtopping due to unexpected increase in inflow to the storage.
Wind break	Type of barrier (vegetative/mechanical) act as fencing wall to protect area from blowing wind
Shelter belt	Rows (more than two) of shrubs of trees which are longer than wind break at angle of 90° to prevailing wind for conservation of moisture and protection of field crops against wind.

3

Hydrology

Technical Term	Definition
Hydrology	Hydrology is the science of water derived from Greek word" hydro + logos" i.e., water + sciences and hance defined as the science which deals with occurrence, distribution and circulation of atmospheric water through unending Hydrologic cycle.
Hydrologic cycle	Hydrologic cycle is defined as the water transfer cycle, in which water is transferred for the ocean/sea to atmosphere them to earth and again to the sea.It Maintain balance b/w water on earth surface & in atmosphere.
Precipitation (PPT)	PPT is all forms of water that reach the earth from atmosphere. OR falling of water on earth surface from atmosphere in forms of rain, show, mist, etc.
Rainfall	Rainfall is the PPT in form of liquid (drop > 0.5 mm) i.e., Liquid PPT is called as Rainfall.
Drizzle	Drizzle is also a liquid form of PPT with drop size <0.5 mm & Intensity < 1mm
Rain gauge	Rain gauge is an instrument used for measuring the depth of PPT at particular point.
Consistency of Rain fall Record	Consistency is determined for testing and adjusting the available rainfall data, especially when condition relevant to rain gauge station has gone significantly change & thus, causing data inconsistent. Double muss curve is used in this method.
Double muss curve	Double muss curve is the graph between cumulative value of rainfall at specified station (expected to have inconsistence data) verses that of from adjoining base stations.

Thiessen polygon method	Method of using a rain gauge network for estimating average depth of rainfall over an area (watershed) and shows areal significance of point rainfall.
Isohyets	Isohyets are the lines joining the points of equal rainfall depth on base map of area. These are referred as contours of PPT.
Rainfall Analysis	Rainfall Analysis is the analytical process of rainfall data collected from various rain gauge stations for determining different hydrological relationships.
Mean Annual Rainfall	The mean annual rainfall is the mean or average of annual rainfall amounts of considered consecutive years.
Losses of PPT	Port of PPT that do generate runoff rather goes in losses in different forms. It is the difference of total rainfall and effective runoff.
Interception (Loss)	Port of PPT or rainfall which intercepted by vegetation foliage, buildings and other object / structure and returned back to atmosphere by evaporation.
Interception Storage	Part of PPT or rainfall which wets the initially dried surfaces of objects (before rainfall event) is called interception storage.
Plant canopy-	Areal vegetative part of plant / tree / crop affecting the penetration & interception of radiant energy.
Depression Storage	Depression Storage is the volume of water retained in the puddles, ditches or some other depressions existing on land surface.
Runoff	Runoff may be defined as the portion of rainfall which flows towards the rivers/stream etc. ORPort of rainfall, which makes its way towards the rivers/stream etc. after satisfying the initial losses.

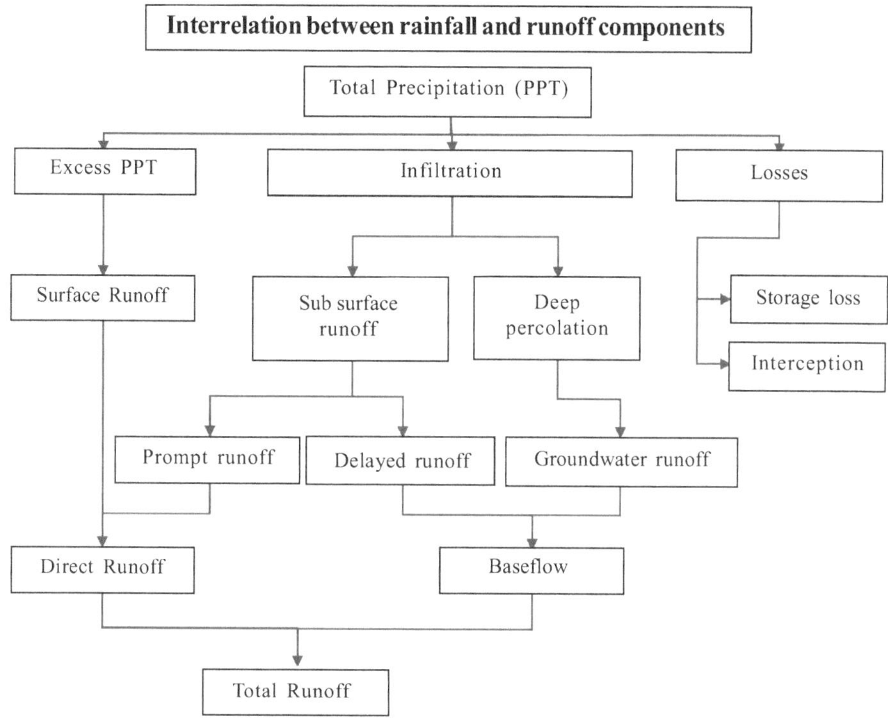

Surface runoff-	The portion of rainfall that reaches the stream without first percolating into the soil.
Surface runoff (Interflow)	The portion of rainfall, which first goes into soil (infiltrate) then starts flowing laterally towards nearby stream/river.
Base flow	The portion of rainfall which first joins the groundwater than flow towards stream/rivers. It is the flow of groundwater towards stream, channel etc. and referred as delayed flow.
Stream flow	When amount of excess rainfall flows through streams (after joining the streams) is called stream flow.
Stream	Flow channel through which runoff from specified basin drains.
Direct runoff	Direct runoff is the sum of surface runoff, interflow and channel precipitation.
Rainfall Distribution	Ratio of maximum rainfall to the mean rainfall of catchment. It presents the effect of rainfall distribution on runoff.

Curve number (CN)	A dimensionless number that relates runoff to the soil-cover complex of a watershed. The higher CN generates higher runoff.
Antecedent Moisture Condition (AMC) of soil	It is the soil moisture condition with antecedent rainfall depth of 5-days. This indicates the wetness status of the watershed before particular evant.
Hydrologic soil group	The classification or group of soils on the basis of their hydrologic characteristics to generate runoff.
™ index	™index is the rate of loss of rain water above which the volume of rainfall is equal to vol of runoff.
W-index	W-index is defined as the average water loss during storm, when rainfall intensity exceeds the Infiltration capacity (max. infiltration rate).
Mass Curve	Curve prepared by plotting cumulative rainfall and time data. Also referred as summation graph.
Double Mass Curve	A curve prepared by plotting the accumulated annual or seasonal precipitation (rainfall) of the specified station (which data is expected to be inconsistent) and concurrent accumulated mean annual precipitation (rainfall) of group of surrounding stations (base stations)
Hyetograph.	It is plot of rainfall intensity and time interval in the form of bar graph.
Rainfall	Rainfall is the PPT in form of liquid (drop > 0.5 mm) i.e., Liquid PPT is called as Rainfall.
Rainfall Excess (Effective Rainfall)	Part of rainfall that exceed the losses and available for runoff. Rainfall that is neither detained on land surfers nor infiltrated into soil.
Design Storm	The specified amount/intensity of storm analyzed by frequency analysis for designing and planning of several structure
Return period	The average number of years within which a given event will be equaled or exceeded
Probability	Likelihood of a certain event to occur
Exceedance Probability	The probability that a random event will exceed a specified magnitude in a given time period, usually one year.
Plotting Position	The paint computed by an equation and used to locate given data on probability paper.

Rainfall Intensity	I = Precipitation / Time
Intensity during relationship	$i = P / T [(T+1) / (t+1)]$
Intensity During Frequency Relationship	$I = KT^a / (t+b)^n$
Depth Area Relationship	$P = P_0 e^{-KA^n}$
Depth Area During Relationship	This relationship is carried out by plotting the progressively decreasing average rainfall depth over progressively increasing watershed area from the center of storm leading to its boundary.
Rainfall Frequency	The frequency of occurrence of rainfall also known as recurrence interval [defined as the period of years during which one storm of a given duration and intensity can be expected to occur].
Probable Maximum Precipitate (Rainfall)	Probable maximum precipitate is defined as the greatest or extreme rainfall for a given duration that is physically possible over a station (basin).
Time of concertation	Time of concertation is the time taken by the surface runoff from hydrologically most remote point of watershed to its outlet.
Hydrograph/Storm Hydrograph/Flood Hydrograph	Hydrograph is a plot (graphical representation) of instantaneous discharge rate against time. It may be called as response of given catchment to a rainfall input
Stream flow Hydrograph	Plot of flow rate (discharge) of stream as function of time at given gauging point of stream. Also called das discharge hydrograph
Annual Hydrograph	It is a plot between stream flow rate and time over a year.
Rising Limb (concentration)	This is an ascending portion of hydrograph curve representing the rising of discharge due to gradual building up storage on surface and channel
Crest Segment	The portion of hydrograph between point of inflection on rising limb and that on falling limb.
Falling Limb (Recession Limb)	This is descending portion of hydrograph curve, extended from end of crest segment (Point of inflection on falling

	limb) up to commencement of base flow (where Surface runoff finished).
Direct Runoff Hydrograph	The hydrograph obtained after base-flow separation is called direct runoff hydrograph
Effective Rainfall Hydrograph	The Hydrograph obtained after subtraction of initial loss and infiltration losses is called Hydrograph
Triangular Hydrograph	It is the plot of rainfall intensity and time plotted in shape of triangle.
Unit Hydrograph	Unit Hydrograph is a direct runoff hydrograph resulting from unit depth of excess rainfall occurring uniformly over basin area at a uniform rate for specified duration
Time Invariant	Direct runoff hydrograph of given effective rainfall in given watershed is always same; time of rainfall occurrence has no effect.
Linear Response/ Principle of Linearity (Superposition)	Relationship between rainfall excess (effective rainfall) and direct runoff is linear.
Method of Superposition	It is a method to develop an unit Hydrograph of n. D-h (where n- integer) from a given D-h unit hydrograph
Method of S-curve	It is a method to develop hydrograph of mD-h (where m fraction or integer) from a given D-h unit hydrograph.
Distribution Graph	It is a D-h unit hydrograph with ordinates showing 6 the percentage of the surface runoff occurring in successive periods of equal time intervals of D-h.
Synthetic unit hydrograph	A unit hydrograph of ungauged watershed, derived based on the empirical relationships relating the basin characteristics to parameters of unit hydrograph is called Synthetic unit hydrograph
Basin lag	It is time interval (h) between the mid-point of unit rainfall excess and peek of unit hydrograph.
Time of concentration	Time taken by surface runoff from hydrological most remote point of watershed to its outlet.
Time Base	It is duration of direct runoff flow.
SCS Dimension Hydrograph	It is a kind of synthetic (unit hydrograph developed by US SCS 1972), such that ordinates of this unit hydrograph are expressed as the ratio of discharge to peak discharge

	(Q/Qp) & abscissa as ratio of time (t) to time to peak (tp).
Instantaneous Unit Hydrograph (IUH)	The IUH is a fictitious/ conceptual unit hydrograph representing the direct runoff hydrograph resulting from an instantaneous PPT of 1 cm effective rainfall of infinitesimally small duration in the watershed. it is a single peak hydrograph with finite base width.
Geomorphological IUH	It is the instantaneous unit hydrograph derived on the basis of geomorphological parameters of watershed.
Stream Gauging	Stream gauging is measurement of discharge /stream flow through stream section in unit time.
Gauging Site (Gauging Point)	The point (location) at stream, on which flow measurement is taken for record is called gauging site
Flood	An unusual high stage in a river overflows its bank and inundates the adjoining areas
Flood Routing.	Prediction of outflow hydrograph for any station at downstream side corresponding to the known inflow hydrograph of one or more station toward upstream side
Reservoir Routing (Storage Routing)	Flood routing through a reservoir (storage) is called Reservoir Routing (Storage Routing)
Attenuation	In reservoir routing there is reduction in peak value of inflow hydrograph due to storage effect this redacting in peak is called attenuation.
Lag	The peak of outflow occurs after the peak of inflow this time difference between two peaks is known as lag
Prism storage	It is the water storage formed between an imaginary line drawn parallel to stream bed or the channel bed
Wedge storage	It is the water storage formed between top of the water profile and imaginary line drawn parallel to streambed

4

Irrigation

Technical Term	Definition
Infiltration	Infiltration is movement of water from the surface into the soil or ground.
Infiltration Rate (velocity)	The actual rate at which water is entering the soil at given time is called infiltration rate / velocity.
Infiltration Capacity	Infiltration Capacity is the maximum value of infiltration rate at the start of infiltration process.
Basic Infiltration rate	The nearly constant rate achieved after some time from start of irrigation is called basic infiltration rate.
Accumulated infiltration	The total quantity of water that infiltrated into the soil at given time is called accumulated/cumulative infiltration.
Surface tension	The property of free water surface that tries to minimize its surface area due to cohesiveness of water molecules and resist the external forces.
Capillarity	Capillarity is a phenomenon of surface tension in soil, due to attraction of water into 'hair-like' openings or capillary of substance.
Capillary rise or fall	Rise or fall of water level above or below original water level, in the capillary tube vertically inserted or held in water.
Soil Moisture tension	Soil Moisture tension is measure of tenacity with which water is retained in soil. It is the force per unit area that must be exerted to remove water from a soil matrix.
PF of soil-	PF Is the logarithm to base 10 of numerical (absolute) value of negative pressure of soil moisture in cm (ie metric head) $pf = \log_{10} h$, where h is soil moisture tension in cm.

Total sol water Potential	Amount of work that must be done at per unit of pure water in order to transport reversibly and isothermally from a pool of pure water (at reference elevation and specific atmospheric pressure) to soil-water at the point under consideration. It is total energy stored in soil-water with respect to referenced condition.
Gravitation Potential	Gravitation Energy stored in per unit quantity of soil water due to difference in elevation with respect to reference point (Both point identical in all respects except elevation).
Pressure Potential	Pressure energy is stored in soil water due to overlying water above point of consideration.
Metric Potential	It is the negative of pressure potential resulting from the capillary and adsorption forces emanating from soil matrix. It refers to energy by which the water is held by soil matrix. It is also called matric suction, capillary potential, soil water suction.
Osmotic Potential (solute Potential)	Energy stored (amount of work) in soil water when it moves to another system identical in all respect except no soluble salt.
Soil moisture stress	Soil moisture stress is referred to or quantify as sumation of soil moisture tension and osmotic potential
osmatic pressure	Pressure difference that must be applied across membrane (permeable only to the solvent) to prevent a net flux of solvent through membrane.
Soil moisture characteristic curve	are curves representing the "available moisture %" as function of "soil moisture tension (atm)" for diffrent soils.
Soil moisture constant	Relatively constant soil moistures which are of some particular significance in agriculture are called soil moisture constant.
Saturation capacity	When all pores are filled with water, soil is said to be at saturation capacity. It is the maximum water holding capacity of soil.
Field Capacity (FC)	Field capacity is soil moisture content when all the gravitational water has drained out of the soil matrix.
Moisture equivalent	Soil moisture content ((moisture retained in soil) of Initially saturated soil after subjects to centrifugal force of 1000 times that of gravity for 30 min.

Permanent willing point (PWP)	Soil moisture content at which plant can no longer obtain enough moisture to meet respiration requirements and remain wilted unless water is added to soil.
Ultimate willing point	Soil moisture content at which plant die (i.e., wilting is completed)
Available water	Amount of soil moisture between FC and PWP and It is the soil moisture which is available for plant use.
Wilting range	Wilting range is range of soil moisture content through which plant undergo progressive degrees of permanent /irreversible wilting.
Readily Available Water (RAW)	Available water is the quantity of water between field capacity and critical soil moisture content.
Maximum available moisture deficiency (MAD)	It the ratio of RAW to available water, helps to estimate amount of water to be used without adversely affecting plant.
Critical soil moisture content	It is soil moisture between FC & PWP such that depletion in soil moisture below this will significantly affects yield of crop.
Water intake	Water intake refers to the movement of water from soil surface into and through the soil. (It is expression of several factors, including infiltration and percolation.)
Percolation	Downward movement of water through saturated on nearly saturated soil in response to force of gravity.
Interflow	Interflow is lateral seepage of water in relatively pervious soil layer above the less pervious layer, generally reappear on surface & Join streams.
Seepage	Seepage is downward and lateral movements of water into sail from source of supply (Reservoir, canal etc.).
Permeability	Characteristic of a pervious medium to permit the fluid to flow through it. It depends only on medium properties
Darcy's Low-	It states that the rate of flow through saturated media is proportional to the hydraulic gradient and cross-section area.
Hydraulic conductivity (k)	Hydraulic conductivity is a proportionality factor in Darcy's law, it is the effective flow velocity at unit hydraulic gradient. It depends on properties of fluid as well as medium both.

Hydraulic head	Hydraulic head is the total energy per unit weight of water, it includes elevation head, pressure head and velocity head.
Hydraulic gradient	Hydraulic head is rate of change of hydraulic head with respect to distance.
Hydraulic equilibrium	The condition of zero flow rate of fluid or film water, when there is net zero hydraulic gradient.
Capillary conductivity	Capillary conductivity is referred to the movement of soil moisture under unsaturated conduit. It also termed as capillary movement and hence $K_{unsaturated}$ is often termed as "capillary conductivity"
Vapor movement	In unsaturated soils, movement of soil water involves both liquid and vapor phase, this movements of vapor is called vapor movement. It predominantly occurs when the metric potential becomes very high which ceases the liquid flow (movement) due to break in continuity, but water moves only in vapor form.
Preferential flow	In natural soil, the part of the infiltrating water travelling faster than the average wetting front. This phenomenon is referred as preferential flow.
Tensiometer	Device with porous cup used to measure soil moisture suction thus available soil moisture indirectly by using soil moisture characteristic curve
Evaporation	Evaporation is a process through which liquid phase convents into gaseous phase.
Transpiration	Transpiration is a process by which of water vapor leaves the living plant body and enters into the atmosphere.
Evapotranspiration (ET)	Denotes the quantity of water evaporated from surface of soil and plant plus water transpired by plant during their growth.
Water requirement of crop	Water needed to meet water losses through ET of a disease-free crop under non-restricting soil condition for achieving full potential under given climate condition, and other unavoidable losses and special applications (if needed).
Potential ET (ET_0)	Evapotranspiration from a large, fully grown healthy vegetation covering land surface with adequate

	moisture all the time is called potential ET. It is considered to be upper limit of ET for a crop in given climate.
Reference ET	Reference evapotranspiration is ET from reference surface (a hypothetical reference crop with an assured crop height of 0.12 m, a fixed surface resistance of 70 S per m^{-1} and an albedo of 0.23) of actively growing grass disease free and with unlimited moisture availability.
Crop ET under standard condition (ET_c)	Evapotranspiration from a given (specified) crop under standard condition (Disease free, well fertilized, fully grown, adequate moisture) with prevailing climate is termed as crop ET under standard condition (ET_c)
Crop ET under adjusted condition (ET_{cadj})	Evapotranspiration from given crop under given conditions (Pests and disease, fertility, growth stage, limited waiter) with given climate is termed as crop ET under adjusted condition (ET_{cadj}).
Irrigation requirement (IR)	Irrigation requirement of a crop refers to the water requirement (WR) of crop, excluding effective rainfall (ER) and soil moisture contribution to plant need (S). IR=WR - (ER) – S
Net irrigation Requirement	Amount of water required to bring soil moisture back to the level of FC in the rootzone soil. (OR) Difference b/w field capacity and soil moisture content of rootzone before irrigation. It does not include any losses.
Gross Irrigation	Total amount of water applied through imitation, including the losses during irrigation.GIR = Net Irrigation Requirement/field efficiency of system
Effective Rainfall	(From agriculture point of view) The portion of rainfall which directly Satisfies crop water needs or stored in crop rootzone.
Irrigation Frequency	It is the days between two successive irrigations during the time when there is no rainfall event.
Irrigation Period	The number of days during which the irrigation is to be applied necessarily if there is no rainfall.
Irrigation scheduling	It is the process of determining the time to irrigate the field and amount of water is to be applied in each irrigation.

Full Irrigation	Full irrigation provides adequate water to meet entire irrigation requirement and aim to get maximum potential production.
Excess Irrigation	Excess irrigation providing water more than the amount that of full irrigation, may reduce yield by adverse effect.
Deficit Irrigation	Deficit Irrigation meets the crop water requirement partially. It is economically justified as when reduction in irrigation below full supply declines the production cost faster. than the reduction in harvested crop value.
Irrigation Efficiency	It is the ratio of water volume used for consumptive use (CU) plus Leaching Requirement (LR) to the water volume supplies to field or pumped for irrigation. Ej = CU + LR / Supply (S) = (S-P-R-O-S) / SEi = Er * Ec * Ea
Reservoir Storge Efficiency (Er)	It is the ratio of volume of water diverted (outflow) from reservoir to the volume of water inflow to the reservoir Er = diverted / inflow = 1-loss / inflow = (inflow – loss) / inflow
Convergence Efficiency	It is the ratio of water delivered to irrigate the field to water diverted from source $EC = W_f / W_d$
Application Efficiency	It is the ratio of water stored in rootzone to water delivered to the field (water applied) Ea = Ws /Wf
Water Storage Efficiency	It is the ratio of water stored in rootzone to water needed in rootzone prior to irrigation Es = Ws / Wn
Water Distribution Efficiency (E_d)	It is the ratio between difference of average depth of water stored along run (d) and average deviation (y, numerical) from average depth stored to the average depth stored along the run (d) Ed = d – y / d = (1- y / d)
Distribution uniformity	It is the ratio of average of low quarter depths of stored water to the average of depth of stored water along the run (d) = $d_{\text{lower quarto valves}}$ / d

Water use efficiency	Water use efficiency denotes the production per unit of water applied i.e. expressed as crop production (kg/hectare) per unit depth of water applied (kg/hectare-cm)
Crop water use Efficiency	Crop water use efficiency is the ratio of crop yield (Y) to amount of water depleted by crop evapotranspiration Crop water use efficiency = Y / ET
Field water use Efficiency	Field water use efficiency is the ratio of crop yield (Y) to total amount of water used in the field (WR)Field water use efficiency = Y / WR
Economic Efficiency of Irrigation	Economic efficiency of irrigation is the ratio of total production (net or gross) profit attained with operating irrigation system, to the total production profit expected from ideal condition. It is a measure of overall efficiency.
Harvest Index	Harvest index is the ratio between total dry matter and harvested crop yield.
Yield Response factor of Crop (k_y)	Yield response factor of crop relates relative yield decrease (1- Ya / Ym) to relative ET deficit (1 - ET a / ET m). (1-Yactual yield / Ymax (potential) = k_y (1-ETactual / ETmax potential) figure 4.1.

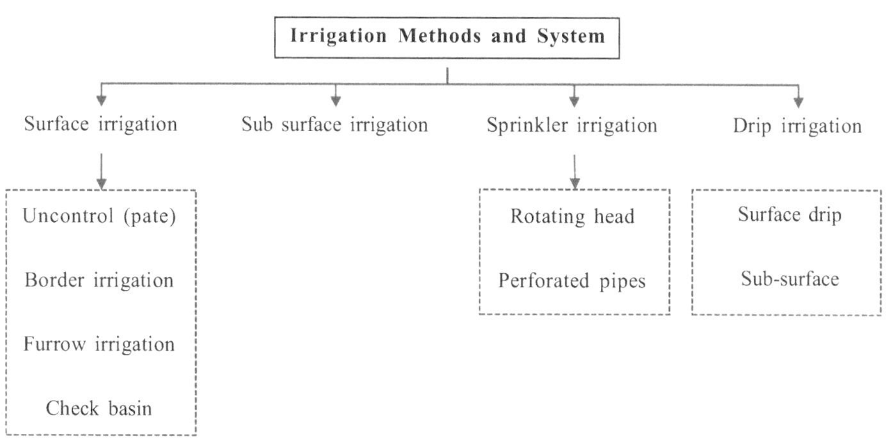

Surface irrigation	It is the method of irrigation in which water is directly applied to soil surface from channel located at the upper reach of field.
Uncontrolled flooding/ (spate) irrigation/ free flooding	The surface irrigation method in which land (field) is irrigated by uncontrolled, freely flowing water.
Border Irrigation	In border irrigation water is applied in form of sheet of flowing water guided by parallel ridges (borders) moves down the slope.
Advance flow/ Front	It is the front of water moving downward when the stream is open for irrigation of border.
Advance Curve	It is the curve between elapsed time and distance of advance front from upstream is called as advance curve.
Recession Flow	It is the flow of tail water receding downstream after cut off of irrigation stream.
Recession Curve	It is the curve between elapsed time from start of irrigation and distance of recession front is called as recession cure.
Time of Ponding	It is vertical distance between recession curve and advance curve at particular (given) distance along the run. (or) The difference between the time at which water (advance) front reaches a particular point along the border and the time at which tail water recodes from same point.
Furrow Irrigation	It is the method of irrigation in which water is applied by running small stream in furrows between the crop rows to irrigate row crops.
Cut back stream (in furrow)	It is the technique of furrow irrigation to obtain optimal opportunity time in which furrow system is to allow the initial design stream to flow full length of furrow and then reduced thestream size.
Currugation Irrigation	It is the irrigation with running water in small furrows (called currugations) flowing down the slope to irrigate close growing crops (small grains, pasture)
Maximum non erosive stream	The maximum size of irrigation stream that can be used at the start of irrigation which is limited by considering erosion in furrow

Check Basin	It is the method of irrigation in which leveled or nearly leveled plots are prepared, surrounded by dikes and allow to be flooded to perform irrigation.
Quarter time rule	It is a thumb rule, used in check basin design, give as "The water spread in entire basin should be covered in one fourth (quarter), time required to infiltrate the net depth of irrigation water.
Sub Surface Irrigation	Method of irrigation in which water is applied below the ground surface by maintaining an artificial water table at some depth (as per soil texture & rootzone.) in this method water reaches into the rootzone through capillary action.
Sprinkler irrigation	Method of irrigation in which water is applied over land surface by spraying it under pressure in form of rain. Often done by rotating head with nozzle and through perforated pipe. Type- 1. Rotating heard- I. Portable II. Semi-portable III. Semi-permanent IV. Solid set V. Permanent system 2. Perforated Pipe
Rotating head sprinkler system	Sprinkler system fitted with rotating heads on riser pipe fixed at regular (uniform) interval along lateral pipe.
Portable sprinkler system	Portable sprinkler system has portable main lines, laterals, and pumping plant, designed to move field to field.
Semi Portable	Similar to (fully) portable system except pumping plant is firm at a location is referred as semi portable
Semi-permanent	System having portable lateral lines but permanent main, sub main and pumping set (plant) is referred as semi-portable sprinkler system.
Solid set	System having enough laterals to eliminate their movement hence lateral are placed in field before season and removed after such sprinkler system is referred as solid set system.
Permanent	A fully permanent system consists of permanently laid (buried)"Mains, sub mains and laterals" and water source and pumping set

Perforated Pipe	Sprinkler system consisting with holes perforated in the lateral pipe in specially designed pattern for uniform distribution of water.
Center pivot system	Sprinkler system in which water source is in center, and a system of pines & sprinkler heads rotates or pivot around center to irrigate field in circular area (shape).
Sprinkler System component	Pumping set, water source, main, lateral, riser, Take off valve, flow control valve, Debris Screen Desalting basin, Booster pump
Takeoff valve	Takeoff valves needed to control pressures in the lateral lines. Used when there is significant pressure difference in main and various lateral take-off points
flow control Valves	Values Used to regulate the pressure and discharge of individual sprinkler (when unequal pressure distribution along lateral occurs).
Debris screen	Usually needed when surface water is used as source and to keep the system free from trash (to avoid sprinkler clogging).
Desilting basin	Usually required to trap sand or suspended silt when water comes from open channels, ditches, or well having sand silt.
Booster Pump	Should be used with existing pumping system when the pump is insufficient to generate required pressure through sprinkler system.
UniformityCoeffi.	Measure of (index of) degree of uniformity of sprinkler system. It is the ratio of difference of average depth of irrigation and average of numerical deviation of irrigation depth from average depth to the average depth of irrigation
Break Up Jet. index	It may be considered as a measure of pressure utilization for breaking up the jet into droplets (diff size) (breakup of jet is necessary for uniform coverage and lessen strike pressure of drop on soil.
Rate of application	Average rate of application. (or PPT intensity) for single sprinkler is rate by which it discharges water.
Fertigation	The method of fertilizer application through pressurized Irrigation systems is called fertigation.

Drip Irrigation (trickle irrigation (Micro irrigation)	Method of irrigation in which water applied frequently with low application rate near to roots of plant through drippers drop by drop (at approx. zero pressure). Localized irrigation
Emitter (water emitter)	Emitters are devices of any kind, type or size, fitted on a pine, are operated under pressure to discharge water (in any form jet, spray mist, drops, small stream or fountain). Sprinkler, Drippers, Spitters, Sprayers, Bubbler, Pulsaters
Drippers	Drippers are small emitters mounted (built in) on irrigation pine line (laterals) at desired spacing in which water inters at certain operating pressure and is discharged at zero. Pressure in forms of chops (rate). Small emitters mounted on (built in) irrigation pipeline to discharge water in forms of drops. ***Note:*** Sprinklers & rain guns - shoot water jets into air Sprayers = Small spray or mist Dripper = Drop by drop (continuous) Bubbler = Small stream or fountain Micro sprinkler = fine droplets distributed by difference.
On and Off trickling	It is a kind of drip irrigation practice in sandy loam soil fields to reduce the downward movement of water and thus moisture availability to crop is increased (on off trickling of 1to2hr)
Sub soil injector (Root injector)	Also called root injector is a sub soil irrigation by injecting water directly to the equipment used for rootzone of plant at depth 15-20 cm from ground surface.

5

Groundwater, Wells and Pumps

Technical Term	Definition
Groundwater	Water present in the zone of saturation (below water table) is called groundwater
Zone of Saturation (Groundwater Zone)	The zone/space in which all the pores are filled with water -The saturated zone over impervious stratum is known as zone of saturation
Well	The vertical hydraulic holes tapped into saturated zone blow ground surface to withdraw water for different purposes.
Water table	Upper surface of zone of saturation is known as water table where water is at atm pressure OR free with surface in an unconfined acquire.
Zone of aeration	The zone / space above the zone of saturation. The zone in which pores are partially filled with water.The space / zone between land surface and water table.
Soil Water Zone	Upper belt of zone of aeration close to land surface where root develops is defined as soil water zone
Capillary Fringe	Zone where water is held by capillary action
Intermediate Zone	Sub zone between soil water belt of capillary fringe
Aquifer	The saturated formation which holds and yield significant quantity of water.
Aquitard	The formation which holds water but does not yield significantly quantity of water. In general, it retards the yield.
Aquiclude	The formation which holds water but impermeable to flow of water. They don't yield.
Aquifuge	The geological formation which neither store (hold) nor yield water (neither porous nor permeable).

unconfined aquifer/ water table aquifer (non artesian)	The aquifer which is not confined by an upper impermeable layer and in which the free water table exists.
Confined aquifer (artesian aquifer)	The aquifer which is confined by an overlying impermeable layer. They have water under pressure more than atmospheric pressure.The aquifer which is confined (bounded) between two impervious beds (layers) aquicludes or aquifuge
Semiconfined	The aquifer which is bounded by semipermeable layer above and permeable or semipermeable layer below is called semiconfined aquifer.
Perched Water table	It is a special type of unconfined aquifer with localized water body above main water table due to locally relatively impermeable stratum.Upper surface of groundwater in peached aquifer is called perched water table.
Saturated Thickness	For confined aquifers, the saturated thickness, H, is equal to the physical thickness of the aquifer between the aquicludes above and below it
Hydraulic Conductivity	The hydraulic conductivity is the constant of proportionality in Darcy's law and is defined as the volume of water that will move through a porous medium in unit time under a unit hydraulic gradient through a unit area measured at right angles to the direction of flow
Transmissibility	Flow through vertical strip extended through full saturated thickness per unit width of aquifer per unit hydraulic gradient $T = Q / w.t = kb$
Drainable Pore Space (Specific yield)	The drainable pore space is the volume of water that an unconfined aquifer releases from storage per unit surface area of aquifer per unit decline of the water table. Drainable pore space is sometimes called specific yield, drainable porosity, or effective porosity.
Storativity	The storativity of a saturated confined aquifer is the volume of water released from storage per unit surface area of aquifer per unit decline in the component of hydraulic head normal to that surface

Yield (Well Yield)	Volume of water pumped per unit time i.e. discharge from a well
Specific Capacity of well	Yield or discharge per unit drawdown.
Open well	Wells dogged out to aquifer near to the ground surface
Tube well	Wells constructed by fixing pine below ground surface passing through different geological formation.
Cavity Well	Type of tube well which doesn't have strainer formed in water weaning send strata lying below a suitable clay layer
Filter point	The tube wells in deltaic region where coarse sand formation is there and are shallow, having well screen and short casing pipe called filter points.
well screen	strainer, separates groundwater from granular material.
Static Water Level	Water level in the well before pumping
Pumping Water Level	Level at which water stands in a well when pumped at given rate
Drawdown	Difference between static water level and pumping water level at that instant
Area of Influence	Area of influence the area which got affected by the pumping of the well
Circle of Influence	Circle of Influence is the boundary of area which gets influenced due to pumping of the well
Radius Influence	Radius of influence is the radius of the circle of influence
Specific Storage	The specific storage of a saturated confined aquifer is the volume of water that a unit volume of the aquifer releases from storage under a unit decline in head.
Hydraulic Resistance	It is the resistance against vertical flow through semi pervious layer of semi-confined aquifer.-Ratio of saturated thickness of semipermeable layer to vertical hydraulic conductivity of flow.
Leakage Factor	The leakage factor describes the spatial distribution of leakage through an aquitard into a semi-confined aquifer, or vice versa.
Piezometers	A piezometer is a small-diameter pipe, driven into, or placed in, the subsoil so that there is no leakage around

	the pipe and all water enters the pipe through its open bottom.
isobath map	depth-to-water table map, or isobath map shows the spatial distribution of the depth of the water table below the land surface
Water table-Fluctuation Map	A water table-fluctuation map is a map that shows the magnitude and spatial distribution of the change in water table over a period
Head-Differences Map	A head-differences map is a map that shows the magnitude and spatial distribution of the differences in hydraulic head between two different soil layers
Well Efficiency	It is a measure of performance of the well.
Well Loss	It is loss in hydraulic head due to resistance offered by well screen to water entering into well.
Well Interference	When the discharge of one well is affected by the pumping of the other nearby well, the well is said to be in condition of well interference. This may typical situation when two wells are located within radius of influence
Pumping Test	Pumping test is a field test (Procedure) to find out the hydraulic characteristics of an aquifer by analyzing the drawdowns of water table in pumping wells & observation wells during pumping.
Well Development	Is the stabilization of walls of well adjacent to well screen (by removing fine particles from vicinity of well screen)
Recuperation test (Recovery test)	Recuperation test is a field test to find out the hydraulic characteristics of an aquifer based on the reduction in drawdown of water table in pumped well after pumping is stopped.
Gravel pack	An artificially graded filter placed immediately around a well screen (to increase the local permeability to prevent small particles to enter into well)
Pump	Pump is a device/machine which transfer energy from external source to fluid (liquid) flowing through conduct (pipe)
Motor	Motor is a machine that convert the hydraulic or electrical energy into mechanical energy

Impeller	The rotating part (of a centrifugal pump or compressor or other machine) designed to move fluid by rotational motion.
Lift Irrigation	Practice of irrigation by lifting the water from lower level with source to irrigate the field at higher level
Flow Irrigation	Practice of irrigation where source at higher level and irrigation is performed by gravity flow
Positive Displacement Pump	Pumps which deliver the constant discharge regardless of pump pressure head
Variable Displacement Pump (Rotodynamic)	Pumps with variable discharge as per the head these have inverse relationship between discharge and head
Centrifugal Pump	Pump in which impeller rotating inside a close fitting case to draw liquid by centrifugal force.
volute Centrifugal Pump	Centrifugal pump in which impeller is surrounding by spiral/ volute type casing
Diffuser Pump (turbine pump)	Centrifugal pump in which Impeller is surrounded by diffuser vanes
Submersible Pump	Vertical turbine pump with close coupled submersible motor
Propeller Pump (axial flow)	Pump in which the flow is parallel to axis head and is mostly developed by propelling or lifting action of impeller
Mixed Flow Pump	Pump which combines the lifting and centrifugal action to generate head by impeller
Jet pump	An assembly of a centrifugal pump (at ground) with an get mechanism in well to lift water.
Airlift Pump	The pump which draws water by injection of compressed air directly into water (through ejector pipe)
Specific Speed	Specific speed is speed in rpm of geometrically similar pump. when delivering discharge (Q) of 1 m^3/s against total head (H) of $1\ mn_s = nQ^{1/2} / H^{/4}$
Pump Characteristic	A graphical description of performance of a pump through curves for head, efficiency, power, NPSH (for centrifugal pump) versus the discharge (x-axis)NPSH: Net Positive Suction Head

Pump efficiency	Hydraulic efficiency of pump expressed as the ratio of energy converted into useful work to energy applied as input.
NPSH (in m)	NPSH is the total suction head at suction nozzle minus the vapor pressure of water at pumping temperature
Affinity Law	Set of equations used to study the effect of diameter and speed on pump characteristics.Set of equations that allow us to alter the performance of a single pump as per need.
Hydraulic Ram	Hydraulic ram is a device used to raise a part of larger amount of water available at some height to a greater height.
Cavitation	Cavitation is a phenomenon of bubbles (gas) formation in liquid (water) typically by the movement of a propeller
Water Hammering (Hydraulic Shock)	Water hammering is a pressure surge (or wave) caused when fluid in motion forced to stop or change direction suddenly.

6

Drainage Engineering

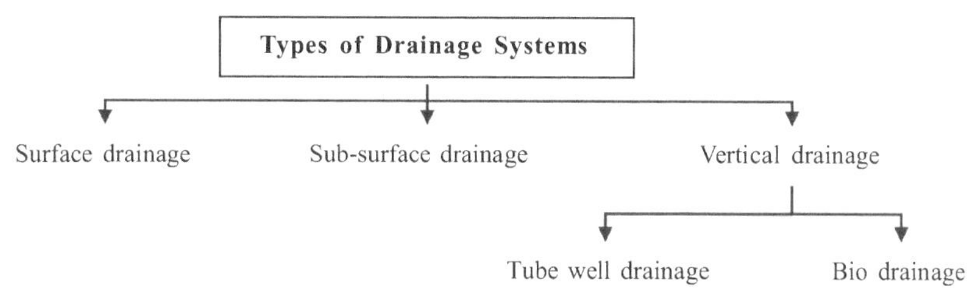

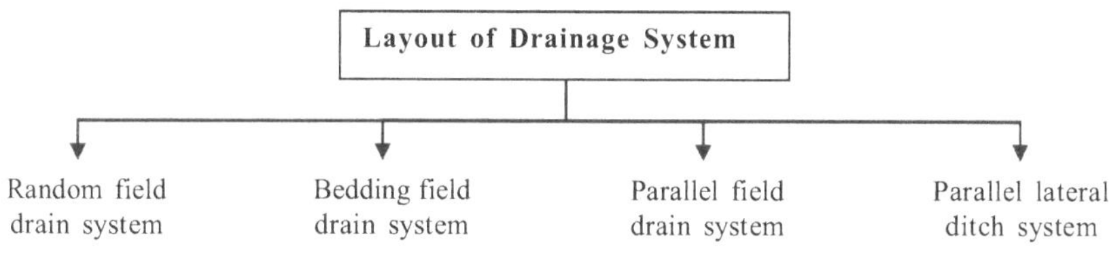

Technical Term	Definition
Drainage	Removal and disposal of excess water from agricultural lands is referred as drainage.
Surface Channels	Removal of excess water from land surface through constructed channels or land shaping.
Surface drainage system	A system of drainage measures i.e. channels and land formingmeant to remove excess surface water.
Bedding sys	It is a method of surface drainage in which land is ploughed into beds separated by dead furrows along slap.
Drainage coefficient	Depth of water to be removed in 24 hr period from entire area, expressed in (cm or depth unit)
Design discharge	A specific value of discharge found out by frequency analysis for designing the dimensions of structure or system or a part thereof.

Interceptor Drain	Drains to control movement of water into drainage problem area also intercept seepage & runoff (from uphill) and drain it away. To protect areas from flooding by surface runoff from adjacent higher ground.
Relief drain	Drains constructed to relief saturated soil of its moisture (parallel to the groundwater flow)
Subsurface Drainage	the process of removal and disposal of excess water present in rootzone below surface.
Subsurface Drainage system	The system of Drainage measures (mole drains, pipe drains) meant to remove excess sub-surface water from root zone
Tile drainage	Subsurface drainage method consists of short length pipes installed at desired depth without and to end joint
Perforated Pipe Drainage	Method of subsurface drainage consists of continuous pipe provided with perforation for entry of drainage water.
Mole Drain	Method of subsurface drainage consist of unlined cylindrical channel at a desired depth in rootzone
Desirable Surplus	Amount of water that must be removed from an area within a certain period of time
Drainage Base	The water level at the outlet of a drained area.
Drainage Criteria	Specified numerical values of drainage parameters that allow a design of drainage system.
Entrance Resistance	The entrance resistance to water flow in the vicinity of a drain pipe due to a decreased permeability and contraction of flow lines
Envelop Material	Material placed around the pipe drains to serve functions like filter, hydraulic and bedding function
Bio Drainage	Removal of excess water through high consuming (with channel of water)
Vertical Drainage	Drainage system where circular vertical wells are used to disposal of excess subsurface water in order to control ground water table.

7

Salt Affected Soils and Management

Technical Term	Definition
Soil	Soil is the superficial unconsolidated and usually weathered part of the mantle of a planet and especially of the earth
Soil Horizons	A soil horizon is defined as a layer of soil or soil material approximately parallel to the land surface and differing from adjacent, genetically-related layers in physical, chemical, and biological properties or characteristics such as color, structure, texture, consistency, and degree of acidity or alkalinity
Soil Profile	The soil profile is defined as a vertical section of the soil, through all its horizons, and extending into the parent material
Anisotropy	Anisotropy means that a substance has different physical properties when measured in different directions.
Soil Texture	The soil consists of primary mineral particles of widely varying sizes. The size distribution of these particles defines the soil's texture (Texture refers to the particle-size distribution of the 'fine earth' of the soil)
Organic Matter	Organic matter is that part of the soil that consists of organic carbon compounds
Soil Consistency	The consistency of the soil refers to the effect of the physical forces of 'cohesion' and 'adhesion' within the soil at various water contents
Soil Structure	The structure of a soil is the binding together of soil particles into aggregates or peds, which are separated from each other by cracks
Soil Fertility	The fertility of a soil, is the ability to supply the nutrients needed by plants for agricultural production

Salt affected soils	The soil having greater amount of accumulated salts them usual which affects the soil plant activities and thus production are called salt affected soils. Categorized as saline, saline-alkali and alkali soils.
Saline Soils	The soils containing soluble salts [chlorides and Sulphates of Na, Mg & Ca] in quantities enough to interfere with crop growth.
Salinization.	The process of accumulation of soluble salt in ln / on soil
Salinity	It is a measure of totally dissolve salts in irrigation water or in soil extract (solution). Expressed as corresponding electric conductivity (EC).
Salt balance Equation	Equation equating the mass (quantity) of salt in different components of mass balance equation i.e. input, output and change in storage for a given soil volume.
Salt equilibrium	It is a situation where there is no long-term change in salt content of rootzone.
Alkali soils (Sodic soil)	The soil having sufficient exchangeable sodium to interface with soil structure and growth of crop.
Alkalinity	Measure (or feature) of extent to indicate the problem of high Sodium content.
Saline alkali soil	The soil which can characterized by EC > 4ds/m, ESP e"15 and pH >8.5.
Cation Exchange capacity	The total quantity of cations which a soil can exchange and adsorb on active surface (negatively changed). It is being expressed as milliequivalent per 100 gm
Dispersed soil	The soil in which clay readily form colloidal suspension with water (in solution) characterized by low permeability, shrinking and cracking when dry and plastic when wet.
Electrical Resistivity	Electrical resistivity is electrical resistance in ohms of a conductor of having 1 cm^2 cross-section area and 1 cm long.
Electrical conductivity	Electrical conductivity is the reciprocal of electrical resistivity
Equivalent weight	Weight of an ion or compound in gm that replace or combines with one gm of hydrogen.Equivalent weight = Atomic weight / valiancy

Exchangeable Sodium Percentage	It is a fraction or percentage of cation exchange capacity (CEC) of soil occupied by sodium ions. ESP = Exchangeable Sodium (m eq /100gm soil)) / CEC (meq / 100gm soil)
Flushing	Flushing is the process of removing salt by surface washing where salts move along excess water. (Suitable for heavy soil)
Leaching	Leaching is the process of dissolving the salt and transporting this dissolved salt by downward movement of water through soil.
Leaching Requirement	Fraction of water entering the soil that must pass through the rootzone in order to prevent soil salinity from exceeding specific value. (To control salinity)
Molar Solution	Solution prepared by dissolving 1 g molecular weight of salt into1 liter of water (gm mol wt / litre)
Soil extract	Solution separated from the soil at particular moisture content.
Saturation extract	Solution separated from the soil at its saturation condition
Sodium adsorption ratio (SAR)	It is a ratio for soil extract and irrigation water used to express the relative activity of sodium ions in exchange reaction with soil SAR = $Na^{+} / [\sqrt{(Ca^{+2} + Mg^{+2})/2}]$;all in m eq/lit
Adjusted SAR	It is a ratio used to classify irrigation water according to the potential to cause infiltration problem because of its high relative sodium content
pH	It is the measure of hydrogen ion concentration expressed as common logarithm (base 10) of the reciprocal of H^{+} ion concentration in mole / liter
Specific ion effect	Any adverse effect of salt constituents on plant growth (as toxic effect to plant, increase osmotic pressure or less (reduced) availability of micronutrients for plants)
Water logging	The accumulation of excess water on the surface or in rootzone. Situation where soils are temporarily saturated or where the groundwater table is to shallow such that capillary rise reaches to rootzone or even soil surface. (Resulting restriction for air simulation & oxygen & increased CO^{2} levels).

Degree of dispersion	It is the ratio of amount of clay dispersed to total clay present in soil.
Gypsum Requirement	Mass of Gypsum (Calcium sulphate) [Caso4 2H,] per unit area that would be required to reduce the exchangeable Sodium percentage of top layer of sodic soil to acceptable level. Gypsum Requirement = 2 (Ca in gypsum - (ca + mg) in filtrate)
Factional salts to be leached	It is given as ration of threshold salinity to initial salinity.
Residual Sodium Carbonate	It is a measure to sodium hazard due to carbonate and bicarbonate of irrigation water, it is generally assumed that all the calcium and magnesium would precipitate as carbonates. Expressed as $RSC = (Co_3^{-2} + H\ CO_3^-) - (Ca^{2+} + Mg^{2+})$
Acidic soils (acidic sulphate soil)	Soils with pH < 4, caused by sulphuric acid formed by oxidation of pyrite (FeS_2)

8

Watershed Geomorphology

Technical Term	Definition
Geomorphological Analysis	Geomorphological analysis is the systematic description of watershed's geometry and its stream channel system to measure the liner aspects of drainage network, aerial aspects of drainage basin and relief aspects of channel network.
Stream order	The stream order represents the degree of stream branching with a watershed
Trunk Stream	The highest order stream through which all runoff water of watershed discharges to outlet is known as trunk stream
Bifurcation ratio	The bifurcation ratio is defined as the ratio of the number of streams of any order to the number of streams of next lower order
Law of stream numbers	Law of stream numbers relates the number of streams of order u^{th} to bifurcation ratio and the principal order
Mean Length of Channel	It is the average length of stream of uth order and given as total length of uth order stream divided by total number of streams of that order
Stream length ratio	Stream length ratio is defined as the ratio of average length of stream of any order to the average length of streams of the next lower order
Law of stream lengths	The law of stream lengths relates the average length of streams of order u to the stream length ratio and the average length of first order streams
Total basin area for u order stream	The area of a basin or order u is the total area projected on a horizontal plane, contributing the overland flow to the streams of given order plus all tributaries of lower order

Law of stream areas	The law of stream area relates the mean area of basin of order u to the mean drainage area of first order stream and the stream area ratio.
Stream area ratio	Stream area ratio is defined as the ratio of average area of stream of any order to the average area of streams of the next lower order
Basin shape	The basin shape is the shape of projected surface on the horizontal plane of basin map
Form factor	Form factor is the ratio of basin area to the square of the basin length
Circulatory ratio	Circulatory ratio is the ratio of basin area to the area of circle having equal perimeter as the perimeter of drainage basin
Elongation ratio	Elongation ratio is the ratio of diameter of a circle having same area as the basin to the maximum basin length
Drainage density	The drainage density is defined as the ratio of the total length of all streams of all order within a watershed to the total area of the watershed
Constant of channel maintenance	The constant of channel maintenance is the inverse of drainage density.
Stream frequency	Stream frequency is the number of stream-segments per unit area of watershed
Relief	It is the elevation difference between reference points located in the drainage basin
Maximum relief	Within the boundary of a basin, the elevation difference between highest and lowest points is known as maximum relief
Maximum basin relief	It is defined as the elevation difference between basin outlet and the highest point located on the perimeter of basin
Relief ratio	It is defined as the ratio of relief to the horizontal distance on which relief was measured
Chanel slope	Chanel slope is the gradient of channel
Stream slope ratio	Stream slope ratio is defined as the ratio of average slope of stream of any order to the average slope of streams of the next lower order

Law of stream slope	The law of stream slopes relates the average slope of streams of order u to the average slope of first order stream and the stream slope ratio
Ruggedness number	The product of relief and drainage density is called ruggedness number
Geometric number	Geometric number is the ratio of ruggedness number to the ground slope
Hypsometric analysis of watershed	The hypsometric analysis of watershed provides the relationship between horizontal cross-sectional area of watershed and the elevation
	Hypsometric curve is plot between relative heights and the relative areas

9

Fluid Mechanics

Technical Term	Definition
Fluid Mechanics	Fluid Mechanics is a study of physical behavior of fluids and fluid systems, and laws governing their behavior. It incorporates action of forces on fluids and the resulting flow pattern
dimension	A dimension is a name which describes the measurable characteristics of an object such as mass, length and temperature etc.
Unit	A unit is accepted standard for measuring the dimension
Density	Density is mass per unit volume= kg/m3
Newton	Newton is that force which when applied to a mass of 1 kg gives an acceleration of 1 m/Sec2. It is a unit of force expressed in terms of mass and acceleration
Pressure	Pressure is the force per unit area (N/ m2 = Pascal or Pa).
Pascal	Pascal is the unit of pressure produced by the force of Newton uniformly applied over an area of 1 m2
Joule	A joule is the work done when the point of application of force of 1 Newton is displaced
Watt	It is a unit of power. A watt represents a work equivalent of a Joule done per second.
Specific weight or Specific density	It is the ratio between the weights of the fluid to its volume. The weight per unit volume of the fluid is called weight density
Specific volume	It is defined as the volume of the fluid occupied by a unit mass or volume per unit mass of fluid is called Specific volume

Specific Gravity	It is defined as the ratio of the Weight density (or density) of a fluid to the Weight density (or density) of a standard fluid
Viscosity	It is defined as the property of a fluid which offers resistance to the movement of one layer of the fluid over another adjacent layer of the fluid.
Kinematic Viscosity	It is defined as the ratio between dynamic viscosity and density of fluid
Newton's law of viscosity	It states that the shear stress (ô) on a fluid element layer is directly proportional to the rate of shear strain
Ideal fluid	A fluid which is compressible and is having no viscosity is known as ideal fluid. It is only an imaginary fluid as all fluids have some viscosity.
Real fluid	A fluid possessing a viscosity is known as real fluid. All fluids in actual practice are real fluids.
Newtonian fluid	A real fluid, in which the stress is directly proportional to the rate of shear strain, is known as Newtonian fluid.
Non-Newtonian fluid	A real fluid in which shear stress is not proportional to the rate of shear strain is known as Non-Newtonian fluid.
Ideal plastic fluid	A fluid, in which shear stress is more than the yield value and shear stress is proportional to the rate of shear strain is known as ideal plastic fluid.
Surface tension	Surface tension is defined as the tensile force acting on the surface of a liquid is contact with a gas or on the surface behaves like a membrane under tension
Capillarity	Capillarity is defined as a phenomenon of rise or fall of a liquid surface in a small tube relative to the adjacent general level of liquid when the tube is held vertically in the liquid.
Vaporizations	A change from the liquid state to the gaseous state is known as Vaporizations.
Absolute pressure	Pressured measured above the absolute zero or complete vacuum is called the absolute pressure
Gauge pressure	Pressure measured above the atmospheric pressure is called Gauge pressure
Vacuum pressure	It is defined as the pressure below the atmospheric pressure

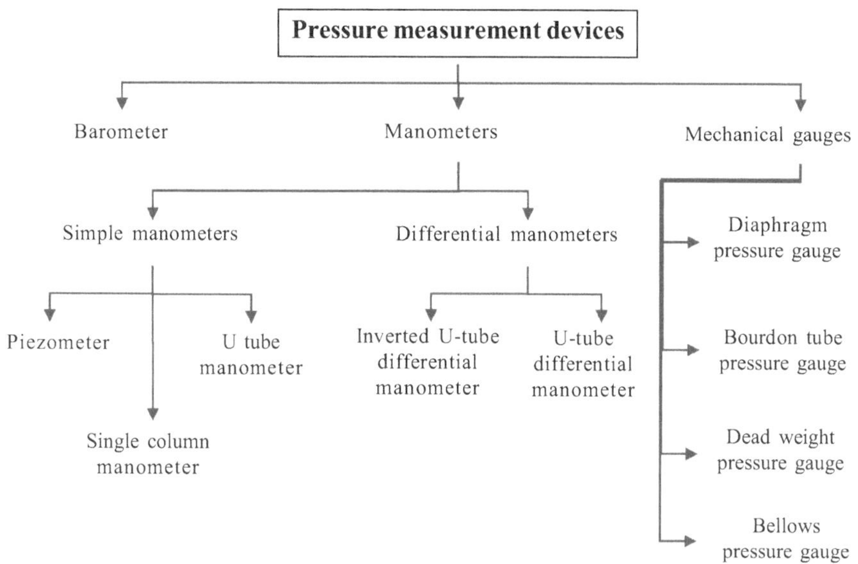

Manometers	Manometers are defined as the devices used for measuring the pressure at a point in a fluid by balancing the column of fluid by the same or another column of fluid
Mechanical Gauges	These are defined as the devices used for measuring the pressure by balancing the fluid column by the spring or dead weight.
Simple manometer	A simple manometer consists of a glass tube having one of its ends connected to a point where pressure is to be measured and the other end remains open to the atmosphere.
Piezometer	It is a simplest form of manometer used for measuring gauge pressure.
U- tube manometer	It consists of a glass tube bent in u-shape, one end of which is connected to a point at which pressure is to be measured and another end remains open to the atmosphere.
Single column manometer	Single column manometer is a modified form of a U-tube manometer in which a reservoir, having a large cross-sectional area as compared to the area of tube is connected to one of the limbs (say left limb) of the manometer.

Differential manometers	Differential manometers are the devices used for measuring the difference of pressure between two points in a pipe or in two different pipes
Kinematics	Kinematics is defined as a branch of science which deals with motion of particles without considering the forces causing the motion
Stream line	A stream line is an imaginary line drawn in a flow field such that the tangent drawn at any point on this line represents the direction of velocity vector
Path line	A path line is locus of a fluid particle as it moves along. In other words, a path line is a curve traced by a single fluid particle during its motion
Streak line.	When a dye is injected in a liquid or smoke in a gas, so as to trace the subsequent motion of fluid particles passing a fixed point, the path fallowed by dye or smoke is called the streak line
Stream tube	If stream lines are drawn through a closed curve, they form a boundary surface across which fluid cannot penetrate. Such a surface bounded by stream lines is known as Stream tube
Steady flow	Steady flow is defined as the flow in which the fluid characteristics like velocity, pressure, density etc. at a point do not change with time.
Un-Steady flow	Un-Steady flow is the flow in which the velocity, pressure, density at a point change with respect to time
Uniform flow	Uniform flow is defined as the flow in which the velocity at any given time does not change with respect to space
Non-uniform flow	Non-uniform is the flow in which the velocity at any given time changes with respect to space
Laminar flow	Laminar flow is defined as the flow in which the fluid particles move along well-defined paths or stream line and all the stream lines are straight and parallel.
Turbulent flow	Turbulent flow is the flow in which the fluid particles move in a zigzag way
Compressible flow	Compressible flow is the flow in which the density of fluid changes from point to point or in other words the density is not constant for the fluid
In-compressible flow	In-compressible flow is the flow in which the density is constant for the fluid flow

Rotational flow	Rotational flow is a type of flow in which the fluid particles while flowing along stream lines also rotate about their own axis
Irrotational flow	If the fluid particles, while flowing along stream lines, do not rotate about their own axis, the flow is called irrotational flow.
1D flow	1D flow is a type of flow in which flow parameter such as velocity is a function of time and one space co-ordinate only, say 'x'
2D flow	2D flow is the type of flow in which the velocity is a function of time and two space co-ordinates, say x & y
3D flow	3D flow is the type of flow in which the velocity is a function of time and three mutually perpendicular directions
Rate of flow or discharge	It is defined as the quantity of a fluid flowing per second through a section of pipe or channel
Continuity equation	The equation based on the principle of conservation of mass is called continuity equation
Fluid dynamics	The study of the forces and energies that are involved in the fluid flow is known as Dynamics of fluid flow
Hydraulic Gradient Line (H.G.L)	If the pressure heads at the different sections of the pipe are plotted to scale as vertical ordinates above the axis of the pipe and all these points are joined by a straight line, a sloping line is obtained, which is known as Hydraulic Gradient Line (H.G.L).
Total energy line (T.E.L).	If at different sections of pipe the total energy (in terms of head) is plotted to scale as vertical ordinate above the assumed datum and all these points are joined, then a straight sloping line will be obtained and is known as energy grade line or total energy line (T.E.L).
Pitot tube	A pitot tube is a simple device used for measuring the velocity of flow.
Venturi meter	A venturi meter is a device used for measuring the rate of flow of fluid through a pipe
Orifice meter	An orifice meter is a simple device for measuring the discharge through pipes

10

Canal Irrigation System and Structures

Technical Term	Definition
Canal Irrigation System	The system in which irrigation water is supplied through irrigation canals network (main canal, submain canal, distributaries, and minors) from source of water (reservoirs) to point of application (agricultural lands)
Alluvial Soils	Alluvial Soil is the soil which is formed by transportation and deposition of silt through the agency of water, over a course of time, is called the alluvial soil
Alluvial Canals	The canals when excavated through alluvial soils, are called alluvial canals
Non-alluvial Soil/area	Mountainous regions may go on disintegrating over a period of time, resulting in the formation of a rocky plain area, called non-alluvial area
Non alluvial Canals	The canals when excavated through Non-alluvial Soil/ area, are called alluvial canals
Watershed Canal or Ridge Canal	The canal which is aligned along any natural watershed (ridge line) is called a watershed canal, or a ridge canal
Contour Canals	The canal which is aligned along contour line is called a contour canal
Side Slope canal	A side slope canal is that which is aligned at rght angles tothe contours; i.e., along the side slopes
Main Canal	Canals constructed on ridge lines to take off water from reservoir to supply the branch canals.
Branch Canals	Branch canals are taken off from the main canal on eather side to take irrigation water for irrigation
Distributaries	Smaller channels which take off from the branch canals and distribute the supply through-outlets into minors-or water courses

Minor	Small channels taken off from the distributaries, so as to supply water to the cultivators at the point nearer to their fields are calle minor
Watercourses	They are small channels, which are excavated and maintained by the cultivators at their own costs, to take water from the government-owned outlet points
Delta (“)	Total quantity of water required by the crop for its full growth
Duty of water (D)	The hectares of land irrigated for full growth of crop by 1m3^3/sec supply of water continuously during entire base period (B) [hectares/cumec]
Base Period (B)	The time between the 1st watering of a crop at time of showing to its last watering before harvesting is called as base period of crop.
Kharif Rabi crop:ratio	The ratio of proposed areas to be ratio (caupratis) in kharif to that at Rabi is called generally 1:2
Paleo irrigation	Irrigation (water application) before first showing of crop particularly in Ravi session.
Kor watering	The first watering which is given to a crop, when crop is of few cm height
Optimum utilization of Irrigation	It means, getting max field with any given amount of water.
Gross Command Area	Total area bounded within the irrigation boundary of a project which can be economically irrigated without considering the limitation of available water quantity. Gross Command Area = Cultivable + Un-cultivable area (Ponds, resident, road)
Cultivable Area	Cultivable area is the part of GCA, on which cultivation is possible. it includes pasture land, fallow land which can be made cultivable
Intensity of irrigation	The percentage of CCA proposed to be irrigated in a given season is called intensity of irrigation
Net shown area	Net shown area is the area which is shown once in a year.
Gross shown area	It is the sum of net shown area and area shown more than once in the same year.

Net irrigated area	The area which is irrigated once in a year.
Gross irrigate area	Sum of net irrigated area in a year and area irrigated more than once in same year.
Area to be irrigated	The area proposed to be irrigated in any season is equals to CCA* Intensity of irrigation
Capacity Factor of Canal	Capacity Factor of Canal is the ratio of mean supply during period to its designed full capacity
Full supply coefficient	Full supply coefficient is design duty at head of canal and given as = Area estimated to be irrigated during base period / Design full supply discharge of head of cancelAlso called Duty on capacity
Nominal duty	Nominal duty is the ratio of area actually irrigated by the cultivators to mean supply discharge let out from outlet of distributary over crop period
Non-agriculture Land	Land occupied by building, roads & railways under water (river, canal, lake etc).
Reporting area for land Utilization Statistics	The area for which data on land use classification are available.
Baran / Uncultivable	The area land under mountains deserts. It is the land which can't be brought under cultivation without very high (exorbitant) cast.
Land Under miscellaneous tree crops	It is the land (cultivable) which not included in 'Net Sown area' but is put to agricultural uses.
Cultivable Wasteland	Land available for cultivation (cultivable land) but not cultivated during last five year or more.
Fallow land	Cultivable land kept fellow (temporarily) for period not less than 1 year & not more than 5 year (between 1 to 5 years).
Current fallow	Cultivable land kept fellow during current year.
Cultivable land	Includes: Net sown area + current fellow + fellow other than current fellow + Cultivable waste + Land under miscellaneous tree cropsNote includes permanent pasture → grassland
Head Works or Diversion Head Works	In any irrigation scheme, a weir or a barrage is constructed across the river, and water is headed up

	on the upstream side, this arrangement is known as head works or diversion head works
Hydraulic Jump	Hydraulic jump is the jump of water that takes place when a super critical flow charges into subcritical flow.
Initial depth (Y1) Sequent depth (Y2)	The depth before the jump is called the initial depth (Y_1) and the depth after the jump is called Sequent depth Y1< Y2
Alternate depth	Depth of flow for which specific energy is the same
Canal Outlet or Module	Canal outlet or a module is a small structure built at the head of the watercourse so as to connect it with a minor or a distributary channel
Non-modular outlets	Non-modular outlets are those through which the discharge depends upon the difference of head between the distributary and the water-course
Semi-modules or Flexible modules	Semi-modules or Flexible modules are those through which the discharge is independent of the water level of the water-course but depends only upon the water level of the distributary so long as a minimum working head is available
Rigid modules or Modular outlets	Rigid modules or Modular outlets are those through which the discharge is constant and fixed within limits, irrespective of the fluctuations of the water levels of either the distributary or of the water course or of both
Cross drainage work	These are the structures constructed at the crossing of an canal and a naturel drain. To avoid crossing canals are general lied along ridge lives

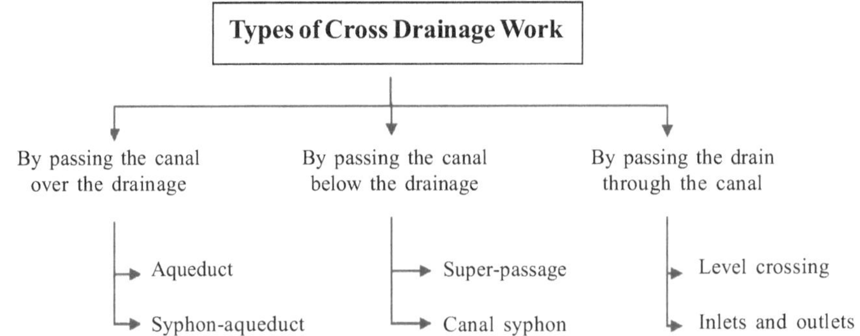

Aqueduct and Syphon Aqueduct	In these works, the canal is taken over the natural drain, such that the drainage water runs below the canal either freely or under syphoning pressure
Super-passage and Canal syphon	In these works, the drain is .taken over the canal such that the canal water runs below the drain either freely or under syphoning pressure
Level Crossing	In this type of cross-drainage work, the canal water and drain water are allowed to intermingle with each other
Dam	A dam may be defined as an obstruction or a barrier built across a stream or a river
Retaining wall	Retaining wall is a construction meant to maintain unequal level of ground surface by its side (back to front) or to Support the soil mass from behind.
Dam	An obstruction or a barrier constructed across the river, stream or valley to retain the water behind for making Lange water body.

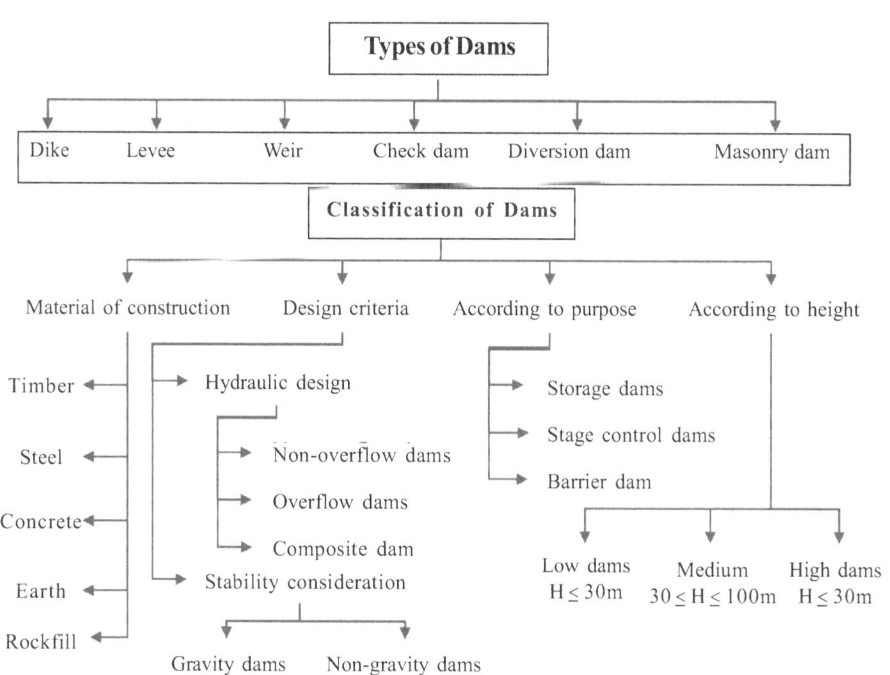

Dike	Dike is a stone or earthen wall constructed as a defense or as a boundary
Reservoir	Reservoir is an artificial lake created by flooding land behind a dam
Spillway	Spillway is a section of a dam designed to pass water from the upstream side of a dam to the downstream side
Levee	Levee is a natural or artificial structure, usually earthen, which parallels the course of a river
Weir	Weir is a small overflow type (designed to be overtopped) dam commonly used to raise the level of a small river or stream
Check dam	Check dam is a small dam designed to reduce flow velocity and control soil erosion
Diversion dam	Diversion dam is a type of dam that diverts all or a portion of the flow of a river from its natural course
Masonry dam	Masonry dam is a type of dam constructed with masonry. It is made watertight by pointing the joints with cement
Coffer dam	Coffer dam is a temporary structure constructed of any material like timber, steel, concrete, rock or earth
Gravity dam	A gravity dam is a solid concrete or masonry structure which ensures stability against all applied loads by its weight alone without depending on arch or beam action
Reservoir sedimentation	The process of sediment deposition in the reservoir refers to the reservoir sedimentation
Sediment	Sediment is a fragmented material, originated from chemical or physical disintegration of rock. Any fragmented material which is transported and deposited by any transporting agent (water, wind, ice)
Sedimentation	The exerted or transported material ultimately getting deposited in water storage bodies called sedimentation.
Sediment load	Total sediment being prospected by agent called sediment load. It is total of saltation, suspension & contact load
Suspended load	Material (Load) moving in suspension from with fluid.

Saltation load	material (Load) moving in from of series of bounces on bed
Contact load (surface creep)	Material (load) moving in rolling & sliding form along bed.
Bed Load	Material (load) moving in contact and near to the bed is called Bed load. It is the form of material moving in saltation and surface creep. [Bed load = saltation load and Contact load]
Farm ponds	Farm ponds are small reservoir like constructions for water harvesting and to solve several purposes of farm needs (irrigation, fishery, cattle feed etc.)
Embankment type Farm Ponds	Embankment type Farm Ponds are the ponds, constructed across the stream or water course by embankment, which dimensions are fixed on basis of volume of water to be stored.
Excavated (dug out) Farm Ponds	Dug-out form ponds are the ponds constructed by excavating the sail from ground, relatively in flat areas (Depth is based on its desired capacity).
Detention ponds	Ponds constructed to hold the run aft, and releasing the same so slowly so that there is no soil erosion of d/s channel
Flood Routing	The procedure that determines the timing and magnitude of a flood wave at a point on a stream from the known or assumed data at one or more points upstream.
Flow Routing	A mathematical procedure that predicts the changing magnitude, speed, and shape of a flood wave as a function of time at one or more points along a watercourse (Maidment 1993)
Reservoir Routing	This procedure derives the outflow hydrograph from a reservoir from the inflow hydrograph into the reservoir with consideration of elevation, storage, and discharge characteristics of the reservoir and spillways
Hydrologic methods of routing	Hydrologic methods are generally based on the solution of the conservation of mass equation and a relation of storage and discharge in a stream reach or reservoir.
Hydraulic methods of routing	Hydraulic methods are based on solutions of the conservation of mass and the conservation of momentum equations.

Translation effect	Translation involves maintaining the same hydrograph shape as the flood wave moves downstream.
Storage or attenuation effect	The storage or attenuation effect involves use of valley storage to reduce the peak flow and change the shape of the hydrograph

11

Land Use Capability Classification

Technical Term	Definition
Land Use Capability	Capability is the inherent capacity of land to perform given at a level for a general use
Land Use Capability Classification	Land Use Capability Classification an exercise of interpretation., grouping and grading of soils according to their potential and limitations for certain use.

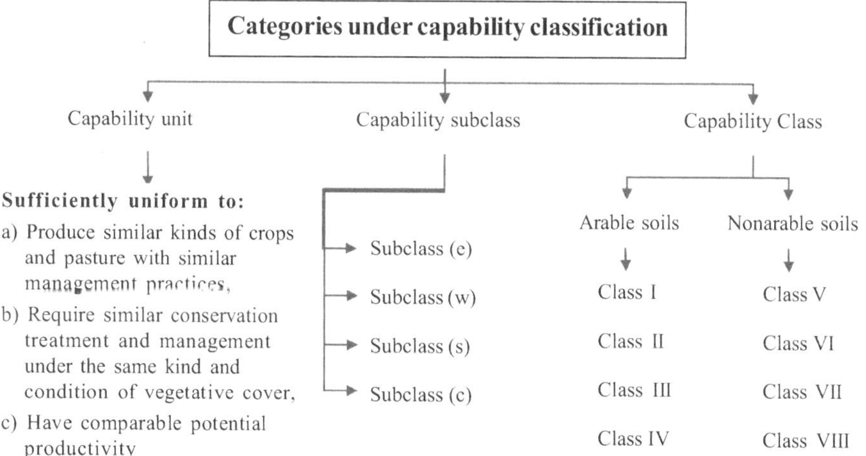

Soil-mapping unit	A soil mapping unit is a portion of the landscape that has similar characteristics and qualities and whose limits are fixed by precise definitions. Within the cartographic limitations and considering the purpose for which the map is made, the soil mapping unit is the unit about which the greatest number of precise statements and predictions can be made.
Capability unit	A capability unit is a grouping of one or more individual soil map- ping units having similar potentials and

continuing limitations or hazards. The soils in a capability unit are sufficiently uniform to (a) produce similar kinds of cultivated crops and pasture plants with similar management practices, (b) require similar conservation treatment and management under the same kind and condition of vegetative cover, (c) have com- parable potential productivity.

Capability unit, is a grouping (subdivision of capability subclass) of soils that have about the same responses to systems of management of common cultivated crops and pasture plants.

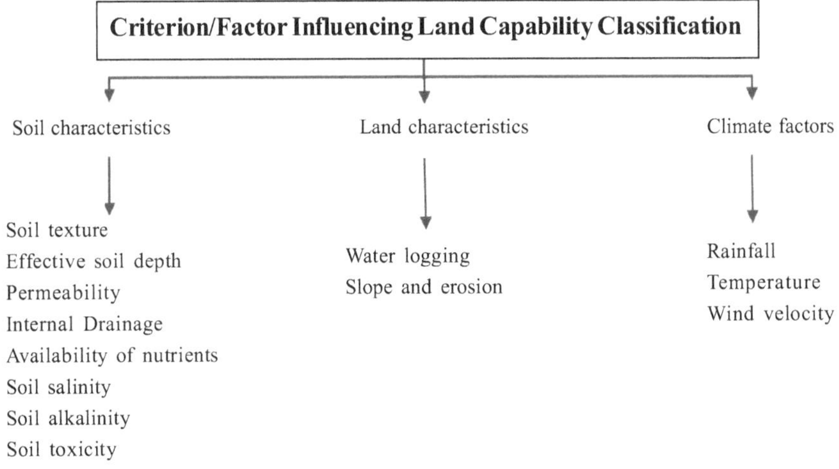

Capability subclass — Subclasses are groups of capability units which have the same major conservation problem, such as-

e-Erosion and runoff

w-Excess water

s-Root-zone limitations

c-Climatic limitations

Capability class — Capability classes are groups of capability subclasses or capability units that have the same relative degree of hazard or limitation. The risks of soil damage or limitation in use become progressively greater from class I to class VIII.

It is the broadest category in the capability classification places all the soils in eight capability classes.

Arable soils

Arable soils are grouped according to their potentialities and limitations for sustained production of the common cultivated crops that do not require specialized site conditioning or site treatment

Colour code of land use capability classification

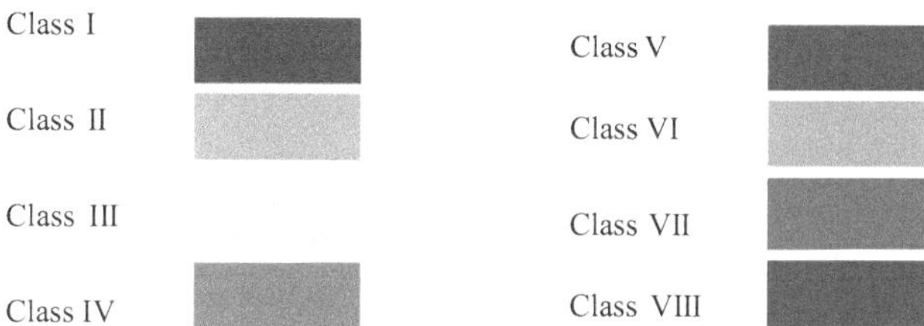

Class I

Soils in this class are suited to a wide range of plants and may be used safely for cultivated crops, pasture, range woodland, and wildlife. The soils are nearly level 6 and erosion hazard (wind or water) is low. They are deep, generally well drained, and easily worked. They hold water well and are either fairly well supplied with plant nutrients or highly responsive to inputs of fertilizer.

Color-Dark Green

Class II

Soils in class II have some limitations that reduce the choice of plants or require moderate conservation practices and may be used for cultivated crops, pasture, range, woodland, or wildlife food and cover.

Color-Pale Green

Class III

Soils in class III have severe limitations that reduce the choice of plants or require special conservation practices, or both. Limitations of soils in class III restrict the amount of clean cultivation; timing of planting, tillage, and harvesting; choice of crops; or some combination of these limitations

Color-Yellow

Class IV

Soils in class IV have very severe limitations that restrict the choice of plants, require very careful management, or both. its use for cultivated crops is limited as a result of the effects of one or more permanent limiting features.

Color-Pink

Nonarable soils	Nonarable soils (soils un- suitable for longtime sustained use for cultivated crops) are grouped ac- cording to their potentialities and limitations for the production of permanent vegetation and according to their risks of soil damage if mismanaged.
Class V	Soils in class V have little or no erosion hazard but have other limitations impractical to remove that limit their use largely to pasture, range, woodland, or wildlife food and cover *Color-Blue*
Class VI	Soils in class VI have severe limitations that make them generally unsuited to cultivation and limit their use largely to pasture or range, woodland, or wildlife food and cover. *Color-Orange*
Class VII	Soils in class VII have very severe limitations that make them unsuited to cultivation and that restrict their use largely lo grazing, woodland, or wildlife *Color-Red*
Class VIII	Soils and landforms in class VIII have limitations that preclude their use for commercial plant production and restrict their use to recreation, wildlife, or water supply or to esthetic purposes. *Color-Purple/Violet*
Class 0	Organic soils (not placed in capability classes)
Subclass (e)	Subclass (e) erosion is made up of soils where the susceptibility to erosion is the dominant problem or hazard in their use. Erosion susceptibility and past erosion damage are the major soil factors for placing soils in this subclass
Subclass (w)	Subclass (w) excess water is made up of soils where excess water is the dominant hazard or limitation in their use. Poor soil drainage, wet- ness, high water table, and overflow are the criteria for determining which soils belong in this subclass
Subclass (s)	Subclass (s) soil limitations within the rooting zone includes, as the name implies, soils that have such

	limitations as shallowness of rooting zones, stones, low moisture-holding capacity, low fertility difficult to correct, and salinity or sodium
Subclass (c)	Subclass(c) climatic limitation is made up of soils where the climate (temperature or lack of moisture) is the only major hazard or limitation in their use

12

Remote Sensing and GIS

Technical Term	Definition
	Geographical Information System
Base Map	Basic representation of a region of the earth as it would appear if viewed from above. That is a map containing geographic features that is used for reference. Roads, for example, are commonly found on base maps
Cartesian Coordinate	Point whose location is expressed in terms of its distance above or below an X, a Y and a Z coordinate plane
azimuth	The horizontal direction of a vector, measured clockwise in degrees of rotation from the positive y-axis, for example, degrees on a compass
Cartography	Cartography is the science and art of making maps and charts.
Digitizing	Digitizing is the process of tracing paper or other documents on a specialized table or "tablet" to capture in digital form lines and points representing map features
Feature	This refers to natural and man-made geographic features represented by points/symbols, lines, and areas on a map. It may also be an object in a geographic or spatial database with a distinct set of characteristics
Geographic Information System (GIS)	Computerized decision support systems that integrate spatially referenced data. These systems capture, store. retrieve, analyze and display spatial data
Global Positioning System (GPS)	Hardware and software designed to communicate with specialized satellites to determine ground location
Ground Control	Ground control refers to points on the surface of the earth with known coordinates as represented by some geographic grid reference system

Map	A map is a representation, usually on a flat (planar) surface, of a region of the earth.
Orthophoto	This term, refers to an aerial photograph that has had distortions due to elevation changes, variation in the distance from the camera to the ground at different locations, and aircraft movement removed
Image Rectification	The process of distortion removal is referred to as image rectification
Overlay	An overlay is something that is laid over or covers something else
Photogrammetry	This term is used in surveying and mapping, and refers to the science, art and technology of obtaining reliable measurements and maps from photographs.
Pixel	This acronym stands for Picture Element, the smallest non-divisible image-forming unit of a plot or video display
Planimetric Map	Horizontal depiction of map features on a two-dimensional plane without any reference to contours or topographic relief
Point	A point is a single X,Y (optionally Z) location in space
Polygon	Closed plane figure bounded by three or more line segments with a nonzero area
Polygon Overlay	A group of polygons on one or more layers, representing various areas that make up a particular geographic theme (e.g., soil types, zoning designations, parcels, land use, etc.)
Raster	Images containing individual dots with color values, called cells (or pixels), arranged in a rectangular, evenly spaced array. Examples: Aerial photographs and satellite images
Rectification	Rectification is a set of techniques for removing data errors though calculation or adjustment
Database management system (DBMS)	A set of computer programs for organizing the information in a database
Datum	A set of parameters and control points used to accurately define the three-dimensional shape of the Earth (e.g., as a spheroid).

Grid	Two sets of parallel lines intersecting at right angles in a plane coordinate system.
Grid cell	A discretely uniform unit that represents a portion of the Earth, such as a square meter or square mile
Image	A graphic representation or description of a scene, typically produced by an optical or electronic device
Map projection	A mathematical model that transforms the locations of features on the Earth's surface to locations on a two-dimensional surface
	Remote Sensing
Remote Sensing	Remote Sensing is the science and technology by which the characteristics of objects of interest can be identified, measured or analyzed the characteristics without direct contact
Passive remote sensing	If the source of the measured energy is the sun, then it is called passive remote sensing
Active remote sensing	If the measured energy is not emitted by the Sun but from the sensor platform then it is defined as active remote sensing
Electromagnetic spectrum	The electromagnetic spectrum is the system that classifies, according to wavelength, all energy (from short cosmic to long radio) that moves, harmonically, at the constant velocity of light
Transmitted	The energy passes through with a change in velocity as determined by the index of refraction for the two media in question.
Absorbed	The energy is given up to the object through electron or molecular reactions.
Reflected	The energy is returned unchanged with the angle of incidence equal to the angle of reflection. Reflectance is the ratio of reflected energy to that incident on a body. The wavelength reflected (not absorbed) determines the color of an object.
Scattered	The direction of energy propagation is randomly changed. Rayleigh and Mie scatter are the two most important types of scatter in the atmosphere.
Emitted	The energy is first absorbed, then re-emitted, usually at longer wavelengths. The object heats up.

Sensors	Device for measuring the electromagnetic radiation at specific ranges (usually called bands) on board of airplanes or on board of satellites
Spatial resolution	It is the resolving power of an instrument needed for the discrimination of features and is based on detector size, focal length, and sensor altitude. Usually measured in pixel size
Spectral resolution	Spectral resolution, is the number and location in the electromagnetic spectrum (defined by two wavelengths) of the spectral bands in multispectral sensors, for each band corresponds an image
Radiometric resolution	Radiometric resolution is the range of available brightness values, which in the image correspond to the maximum range of Digital Numbers. Usually measured in bits (binary digits),
Temporal resolution	temporal resolution is the time required for revisiting the same area of the Earth
Radiance	Radiance corresponds to the brightness in a given direction toward the sensor
Reflectance	The ratio of reflected versus total power energy.
Spectral signature	The spectral signature is the reflectance as a function of wavelength, each material has a unique signature, therefore it can be used for material classification
supervised classification	Supervised classification is an image processing technique that allows for the identification of materials in an image, according to their spectral signatures
Color composite	A combination is created of three individual monochrome images, in which each is assigned a given color; this is defined color composite
Training Areas	Supervised classifications require the user to select one or more Regions of Interest (ROIs, also Training Areas) for each land cover class identified in the image
Image processing	Remote sensing images can be processed in various ways in order to obtain classification, indices, or other derived information that can be useful for land cover characterization

Principal Component Analysis	Principal Component Analysis (PCA) is a method for reducing the dimensions of measured variables (bands) to the principal components
Pan-sharpening	Pan-sharpening is the combination of the spectral information of multispectral bands (MS), which have lower spatial resolution with the spatial resolution of a panchromatic band (PAN)
Spectral Indices	Spectral indices are operations between spectral bands that are useful for extracting information such as vegetation cover
Clustering	Clustering is the grouping of pixels based on spectral similarity calculated for a multispectral image.
Radiance	Radiance is the flux of energy (primarily irradiant or incident energy) per solid angle leaving a unit surface area in a given direction.

Appendix

Book list and ARS Syllabus (L & WME) CODE-53

List of books for further study

Detail Study/ Descriptive

No.	Book Name	Author
1	Irrigation Theory and Practice	Michael A. M.
2	Land and Water Management Engineering	Murty V.V.N. and Jha M.K.
3	Soil and water Conservation Engineering	Suresh R
4	Manual of Soil Water Conservation Practices	Singh G., Venkataraman C., Sastry G. and Joshi B.P.
5	Watershed Hydrology	Suresh R
6	Fluid Mechanics and Hydraulic Machines	Bansal R.K.
7	Water Well and Pump Engineering	A.M. Michael
8	Watershed Planning and Management	R.V, Singh,
9	Drainage Principles and Applications (2 ed.). ILRI Publication	Ritzema H. P.
10	Irrigation Engineering and Hydraulic Structures	Garg S. K
11	Hand Book of Agricultural Engineering	ICAR
12	Principles of Agricultural Engineering. (Vol. II)	Michael, A. M. and O jha, T.P.
13	Engineering Hydrology	Subrahmanya, K
	Books for quick revision/ Objective	
1	Objectives in Soil & Water Conservation Engineering	Pawan Jeet Prem
2	Land and Water Management Engineering	Ravish Keshri and Rao K.V. R.
3	An Objective Review in Soil and Water Engineering	Surseh R.
4	Land and Water Management Engineering	Murty V.V.N. and Jha M.K.
5.	*Objectives Given at end of book chapters*	
	Other Reference Books	
1.	Elementary Hydrology	Singh V. P
2.	Hand Book of Applied Hydrology	Chow V.T

3.	Open Channel Hydraulics	Chow V. T.
4.	Irrigation Water Management: Principles and Practice	Majumdar D. K.
5.	Irrigation Hydraulics	Lal, R
6.	Drainage Engineering.	Luthin, J.
7.	Irrigation and Water Resources Engineering	Asawa G. L.